PALM TREES OF THE AMAZON

AND THEIR USES.

BY

ALFRED RUSSEL WALLACE.

WITH FORTY-EIGHT PLATES.

1853

British Library Cataloguing-in-Publication Data
A catalogue record for this book is available from the
British Library

MAP
of
AMERICA
shewing the Distribution of
PALMS.

Alfred Russel Wallace

Alfred Russel Wallace was born on 8th January 1823 in the village of Llanbadoc, in Monmouthshire, Wales.

At the age of five, Wallace's family moved to Hertford where he later enrolled at Hertford Grammar School. He was educated there until financial difficulties forced his family to withdraw him in 1836. He then boarded with his older brother John before becoming an apprentice to his eldest brother, William, a surveyor. He worked for William for six years until the business declined due to difficult economic conditions.

After a brief period of unemployment, he was hired as a master at the Collegiate School in Leicester to teach drawing, map-making, and surveying. During this time he met the entomologist Henry Bates who inspired Wallace to begin collecting insects. He and bates continued exchanging letters after Wallace left teaching to pursue his surveying career. They corresponded on prominent works of the time such as Charles Darwin's *The Voyage of the Beagle* (1839) and Robert Chamber's *Vestiges of the Natural History of Creation* (1844).

Wallace was inspired by the travelling naturalists of the day and decided to begin his exploration career collecting specimens in the Amazon rainforest. He explored the Rio Negra for four years, making notes on the peoples and languages he encountered as well as the geography, flora, and fauna. On his return voyage his ship, Helen, caught fire and he and the crew were stranded for ten days before being picked up by the Jordeson, a brig travelling from Cuba to London. All of his specimens aboard Helen had been lost.

After a brief stay in England he embarked on a journey to the Malay Archipelago (now Singapore, Malaysia, and Indonesia). During this eight year period he collected more than 126,000 specimens, several thousand of which represented new species to science. While travelling, Wallace refined his thoughts about evolution and in 1858 he outlined his theory of natural selection in an article he sent to Charles Darwin. This was published in the same year along with Darwin's own theory. Wallace eventually published an account of his travels *The Malay Archipelago* in 1869, and it became one of the most popular books of scientific exploration in the 19th century.

Upon his return to England, in 1862, Wallace became a staunch defender of Darwin's landmark work *On the Origin of Species* (1859). He wrote responses to those critical of the theory of natural selection, including 'Remarks on the Rev. S. Haughton's Paper on the Bee's Cell, And on the Origin of Species' (1863) and 'Creation by Law' (1867). The former of these was particularly pleasing to Darwin. Wallace also published important papers such as 'The Origin of Human Races and the Antiquity of Man Deduced from the Theory of 'Natural Selection" (1864) and books, including the much cited *Darwinism* (1889).

Wallace made a huge contribution to the natural sciences and he will continue to be remembered as one of the key figures in the development of evolutionary theory.

Wallace died on 7th November 1913 at the age of 90. He is buried in a small cemetery at Broadstone, Dorset, England.

PREFACE

THE materials for this work were collected during my travels on the Amazon and its tributaries from 1848 to 1852. Though principally occupied with the varied and interesting animal productions of the country, I yet found time to examine and admire the wonders of vegetable life which everywhere abounded. In the vast forests of the Amazon valley, tropical vegetation is to be seen in all its luxuriance. Huge trees with but-tressed stems, tangled climbers of fantastic forms, and strange parasitical plants everywhere meet the admiring gaze of the naturalist fresh from the meadows and heaths of Europe. Everywhere too rise the graceful Palms, true denizens of the tropics, of which they are the most striking and characteristic feature. In the districts which I visited they were everywhere abundant, and I soon became interested in them, from their great variety and beauty of form and the many uses to which they are applied. I first endeavoured to familiarize myself with the aspect of each species and to learn to know it by its native name; but even this was not a very easy matter, for I was often unable to see any difference between trees which the Indians assured me were quite distinct, and had widely different properties and uses. More close examination, however, convinced me that external characters did exist by which every species could be separated from those most nearly allied to it, and I was soon pleased to find that I could distinguish one palm from another, though barely visible above the surrounding forest, almost as certainly as the natives themselves. I then endeavoured to define the peculiarities of form or structure which gave to each its individual character, and made accurate sketches and descriptions to impress them upon my memory. These peculiarities are often very slight, though permanent:— in the roots, the extent to which they appear above the ground;—in the stem, the thickness, which in each species varies within very definite limits,—the swelling of the base, the middle or the summit,—its generally erect or curving position,—the nature of the rings with which it is marked,—the number, direction and form of the spines or tubercles with which it is armed;—in the leaves, the erect or drooping position, the size and form of the leaflets, the angles which they form with the midrib, and the proportionate size of the terminal pair, are all important characters. The fruit spike or spadix is either erect or drooping, either simple, forked, or many-branched; and the fruits in closely allied species vary in size, in shape, and in colour, as well as in the bloom, down, hairs or tubercles with which they are clothed.

In this little work careful engravings from myoriginal drawings are given, with a general description of each species, and a history from personal observation of the various uses to which it is applied, and of any other interesting particulars connected with it. Several of the species here figured are new, and among them is the Palm which produces the "piassába," the coarse fibrous material of which brooms for street-sweeping are now generally made.

For the determination of the genera and species, and for that part of the Introduction relating to the botanical characters and geographical distribution of Palms, I am indebted to the magnificent work of Dr. Martius. To the botanist I trust my little book may be of some use, in giving accurate figures of many entire plants, of which he is only acquainted with small portions, and in supplying an account of the uses to which they are applied in the distant regions where they grow. And to the general reader I hope it may not be uninteresting, as exhibiting a glimpse of a wild and rude people in the lowest state of civilization, whose existence is intimately connected with the products of the surrounding forests, among which the plants under consideration hold so prominent a place; and of these it is hoped the accompanying Plates will give a more accurate idea than the stereotyped figures which often represent the "feathery palm trees" in our popular works.

Some of the fruits of which I had no drawings, have been figured from specimens in the Museum at Kew collected by Mr. R. Spruce, who is still investigating the Botany of the Amazon valley.

London, October 1853.

LIST OF PLATES.

LIST OF PLATES.

Pl. II. PALM FRUITS

1. Raphia tædigera 2. Mauritia flexuosa 3. Manicaria fera 4. Lepidocaryum tonue 5. Astrocaryum tucuma 6. Leopoldinia pulchra

Pl. III. PALM FRUITS

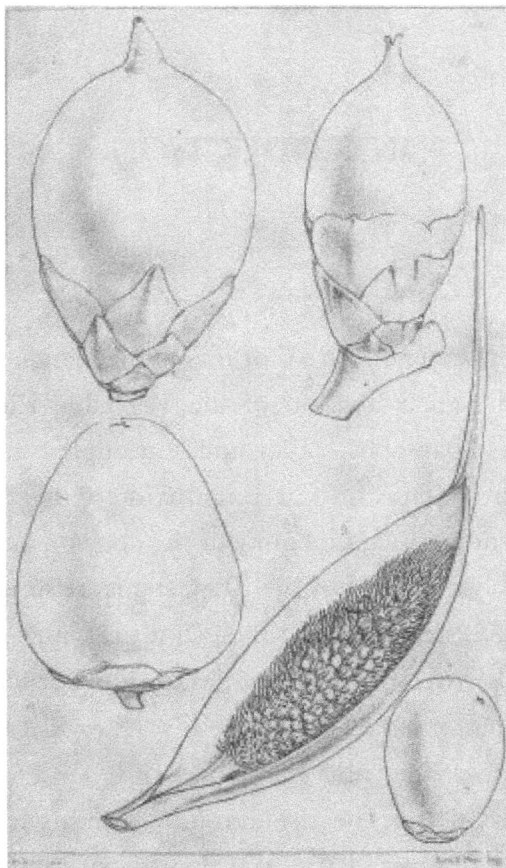

1. Attalea spectabilis 2. Maximiliana regia 3. of Mregia 4.
Guilielma speciosa 5. Iriartea exorhiza

PALM TREES OF THE AMAZON AND THEIR USES.

INTRODUCTION.

PALMS are endogenous or ingrowing plants, belonging to the same great division of the Vegetable Kingdom as the Grasses, Bamboos, Lilies and Pineapples, and not to that which contains all our English forest trees. They are perennial, not annual like most of the above-named plants, and probably reach a great age. Their stems are simple or very rarely forked, slender, erect, and cylindrical, not tapering as in most other trees; they are hardest on the outside, and are marked more or less distinctly with scars or rings, marking the situation of the fallen leaves.

The leaves are generally terminal, forming a bunch or head at the summit of the tree; they are of very large size, have long petioles or footstalks, and are alternately placed on the stem. In shape they are pinnate or flabellate, or rarely simple, sheathing at the base, without stipules; and they have a plicate vernation, or are folded up lengthways before they open. The margins of the sheathing bases of the leaf-stalks are often fibrous, and give out a variety of singular processes.

The flowers are numerous, small, symmetrical, un-coloured, or obscurely so, six-parted, and hermaphrodite or polygamous. They are produced in a spadix from the axils of the leaves, and are generally enclosed in a spathe or sheath. The ovary or seed-vessel is three-celled or three-lobed, but the fruit is generally one-seeded from abortion, and the seed is large and albuminous with a fibrous or fleshy covering.

Palms are almost exclusively tropical plants, very few species being found in the temperate zone, and those only in the warmer parts of it, while the nearer we approach the equator the more numerous they become both in species and individuals. Dr. Martius, a Prussian botanist and traveller in South America, has published a magnificent work in three folio volumes, entirely devoted to the Botanical history of this family of plants. He divides the portion of the earth which produces palms into five regions, namely,—

The North Palm Zone, extending from the northern limit of Palms to the tropic of Cancer.

The transition North Palm Zone, from the tropic of Cancer to 10° north latitude.

The Chief Palm Zone, from 10° north to 10° south latitude.

The transition South Palm Zone, from 10° south latitude to the tropic of Capricorn, and

The South Palm Zone, from the tropic of Capricorn to the southern limit of the family.

The Northern limit of Palms is, in Europe 43° of latitude, in Asia 34°, and in America 34°.

The Southern limit is 34° in Africa, 38° in New Zealand, and 36° in South America.

To the north of the tropic of Cancer there are 43 species of Palms known, and to the south of the tropic of Capricorn only 13, while as we advance from either side towards the equator the number increases, until in the Chief Zone, between 10° north and 10° south latitude, there are more than 300 species (see Frontispiece Map).

In the Old World, the rich islands of the Eastern Archipelago produce the greatest number of Palms; in the New, the great valleys of the Amazon and Orinoco on the main land, are most prolific.

In proportion to its extent, America is the most productive palm country; for while the Old World, including Europe, Asia, Africa, and the Eastern Archipelago, with New Holland and all the Pacific Islands, contain 307 species, the New World or America alone has 275 different kinds.

In the Old World the islands produce more species than the continents, the former containing 194, while the latter have only 113.

In the New World, however, the reverse is the case, the continent there containing 234, while the islands possess only 42 kinds of Palms.

The total number of Palms at present known is less than 600. Dr. Martius thinks that the probable number existing on the earth may be from 1000 to 1200; though, as similar calculations have hitherto almost invariably been proved, as our knowledge increased, to be far below the truth, it is not unlikely that a few years may render double this number a more probable estimate.

Palms present to our view the most graceful and picturesque, as well as some of the most majestic forms in the vegetable kingdom. Though many of them have a sameness of aspect, yet there is a sufficient contrast and variety of forms to render them interesting objects in the landscape. The stems in some species do not appear above the ground, in others they rise to the height of 200 feet; some resemble reeds and are no thicker than a goose quill, others swell out to the bulk of a hogshead. There are climbing palms too, which trail their long flexible stems over trees and shrubs, or hang in tangled festoons between them.

The trunks of some are almost perfectly smooth, others rough with concentric rings, or clothed with a woven or hairy fibrous covering, which binds together the sheathing bases of the fallen leaves. Many are thickly beset with cylindrical or flat spines, often 8 or 10 inches long and as sharp as a needle; and the fallen leaves and stems of these offer a serious obstacle to the traveller who attempts to penetrate the tropical forests.

The leaves are large and often gigantic, surpassing those of any other family of plants. In some species they are 50 feet long and 8 wide; these are pinnate or composed of numerous long narrow leaflets placed at right angles to the midrib, but in others the leaves are entire and undivided, and yet are 30 feet or more in length and 4 or 5 in width. But the most remarkable form of leaf is the fan-shaped, which characterizes a considerable number of species, and gives them such a completely different aspect, as to render it, to ordinary observers, the most palpable feature dividing the whole family into two distinct groups. The Palms having fan-shaped leaves are, however, comparatively few, being only 91 out of 582 known species.

The flowers are small and inconspicuous, generally of a white, pale yellow or green colour, but often produced in such dense masses as to have a striking appearance. They sometimes emit a very powerful odour, which attracts swarms of minute insects; and a newly-burst palm spathe may often be discovered by the buzzing cloud of small flies and beetles which hover over it.

The fruits are generally small, when compared with the size of the trees; the common cocoa-nut being one of the largest in the whole family. The kernel of many is too hard to be eaten, and the outer covering is often fibrous or woody; but in others the seeds are covered with a pulpy or farinaceous mass, which in most cases furnishes a grateful and nutritious food.

The purposes to which the different parts of Palms are applied are very various, the fruit, the leaves, and the stem all having many uses in the different species. Some of them produce valuable articles of export to our own and other countries, but they are of far more value to the natives of the districts where they grow, in manycases furnishing the most important necessaries for existence.

The Cocoa-nut is known to us only as an agreeable fruit, and its fibrous husk supplies us with matting, coir ropes, and stuffing for mattresses; but in its native countries it serves a hundred purposes; food and drink and oil are obtained from its fruit, hats and baskets are made of its fibre, huts are covered with its leaves, and its leaf-stalks are applied to a variety of uses. To us the Date is but an agreeable fruit, but to the Arab it is the very staff of life; men and camels almost live upon it, and on the abundance of the date harvest depends the wealth and almost the existence of many desert tribes. It is truly indigenous to those inhospitable wastes of burning sand, which without it would be uninhabitable by man.

A palm tree of Africa, the *æleis guianensis,* gives us oil and candles. It inhabits those parts of the country where the slave trade is carried on, and it is thought by persons best acquainted with the subject that the extension of the trade in palm oil will be the most effectual check to that inhuman traffic; so that a palm tree may be the means of spreading the blessings of civilization and humanity among the persecuted negro race.

Sago is another product of a palm, which is of comparatively little importance to us, but in the East supplies the daily food of thousands. In many parts of the Indian Archipelago it forms almost the entire subsistence of the people, taking the place of rice in Asia, corn in Europe, and maize and mandiocca in America, and is worthy to be classed with these the most precious gifts of nature to mankind. Unlike them, however, it is neither seed nor root, but is the wood itself, the pithy centre of the stem, requiring scarcely any preparation to fit it for food; and it is so abundant that a single tree often yields six hundred pounds weight.

The canes used for chair bottoms and various other purposes, are the stems of species of *Calamus,* slender palms which abound in the East Indian jungles, climbing over other trees and bushes by the help of the long hooked spines with which their leaves are armed. They sometimes reach the enormous length of 600 or even 1000 feet, and as four millions of them are imported into this country annually, a great number of persons must find employment in cutting them.

A variety of species, in all parts of the world, furnish a sugary sap from their stems or unopened spathes, which when partly fermented is the palm wine of Africa and the Toddy of the East Indies; and a similar beverage is procured from the *Mauritia vinifera* and other species in South America. Indeed, at the mouth of the Orinoco dwell a nation of Indians whose existence depends almost entirely on a species of Palm, supposed to be the *Mauritia flexuosa*. They build their houses elevated on its trunks, and live principally upon its fruit and sap, with fish from the waters around them.

Among the most singular products of palm trees are the resins and wax produced by some species. The fruits of a species of *Calamus* of the Eastern Archipelago are covered with a resinous substance of a red colour, which, in common with a similar product from some other trees, is the Dragon's blood of commerce, and is used as a pigment, for varnish, and in the manufacture of tooth powder. The *Ceroxylon andicola*, a lofty palm growing in the Andes of Bogotá, produces a resinous wax which is secreted in its stem and used by the inhabitants of the country for making candles and for other purposes. Again, in some of the northern provinces of Brazil is found a palm tree called Carnaúba, the *Copernicia cerifera,* having the underside of its leaves covered with white wax, which has no admixture of resin, but is as pure as that procured from our hives.

The leaves of palms, however, are applied to the greatest variety of uses; thatch for houses, umbrellas, hats, baskets and cordage in countless varieties are made from them, and every tropical country possesses some species adapted to these varied purposes, which in temperate zones are generally supplied by a very different class of plants. The Chip, or Brazilian-grass hats, so cheap in this country, are made from the leaves of a palm tree which grows in Cuba, whence they are imported for the purpose: the palm is the *Chamærops argentea;* and in Sicily an allied species, the *Chamærops humilis* (the only European palm), is applied in a similar manner to form hats, baskets, and a variety of useful articles.

The papyrus of the ancient Egyptians, and the metallic plates on which other nations wrote, were not used in India, but their place was supplied by the leaves of palms, on whose hard and glossy surface the characters of the Pali and Sanscrit languages were inscribed with a metallic point. The leaves of the *Corypha taliera* are used for this purpose, and when strung together, form the volumes of a Hindu library.

A favourite stimulant too of the Malays is furnished by a palm. The fruit of the *Areca catechu* is the betelnut, which they chew with lime, and which is their substitute for the opium of the Chinese, the tobacco of Europeans, and the coca of the South Americans.

One of the most recent introductions into our own domestic economy is the fibre of a palm, the Piassaba, which is now generally used for coarse brooms and brushes; and in the valley of the Amazon, of which it is a native, the same material is manufactured into cables, which are cheap and very durable in the water.

We have now glanced at a few of the most important uses to which Palms are applied, but in order to be able to appreciate how much the native tribes of the countries where they most abound are dependent on this noble family of plants, and how they take part in some form or other in almost every action of the Indian's life, we must enter into his hut and inquire into the origin and structure of the various articles we shall see around us.

Suppose then we visit an Indian cottage on the banks of the Rio Negro, a great tributary of the river Amazon in South America. The main supports of the building are trunks of some forest tree of heavy and durable wood, but the light rafters overhead are formed by the straight cylindrical and uniform stems of the Jará palm. The roof is thatched with large triangular leaves, neatly arranged in regular alternate rows, and bound to the rafters with sipós or forest creepers; the leaves are those of the Caraná palm. The door of the house is a framework of thin hard strips of wood neatly thatched over; it is made of the split stems of the Pashiúba palm. In one corner stands a heavy harpoon for catching the cow-fish; it is formed of the black wood of the *Pashiúba barriguda.* By its side is a blowpipe ten or twelve feet long, and a little quiver full of small poisoned arrows hangs up near it; with these the Indian procures birds for food, or

23

for their gay feathers, or even brings down the wild hog or the tapir, and it is from the stem and spines of two species of Palms that they are made. His great bassoon-like musical instruments are made of palm stems; the cloth in which he wraps his most valued feather ornaments is a fibrous palm spathe, and the rude chest in which he keeps his treasures is woven from palm leaves. His hammock, his bow-string and his fishing-line are from the fibres of leaves which he obtains from different palm trees, according to the qualities he requires in them,—the hammock from the Mirití, and the bow-string and fishing-line from the Tucúm. The comb which he wears on his head is ingeniously constructed of the hard bark of a palm, and he makes fish hooks of the spines, or uses them to puncture on his skin the peculiar markings of his tribe. His children are eating the agreeable red and yellow fruit of the Pupunha or peach palm, and from that of the Assaí he has prepared a favourite drink, which he offers you to taste. That carefully suspended gourd contains oil, which he has extracted from the fruit of another species; and that long elastic plaited cylinder used for squeezing dry the mandiocca pulp to make his bread, is made of the bark of one of the singular climbing palms, which alone can resist for a considerable time the action of the poisonous juice. In each of these cases a species is selected better adapted than the rest for the peculiar purpose to which it is applied, and often having several different uses which no other plant can serve as well, so that some little idea may be formed of how important to the South American Indian must be these noble trees, which supply so many daily wants, giving him his house, his food, and his weapons.

To the lover of nature Palms offer a constant source of interest, reminding him that he is amidst the luxuriant vegetation of the tropics, and offering to him the realization of whatever wild and beautiful ideas he has from childhood associated with their name.

In the equatorial regions of South America they are seldom absent. Either delicate species flourishing in the dense shade of the virgin forest; or lofty and massive, standing erect on the river's banks; or on the hill side raising their leafy crowns on airy stems above the surrounding trees, creating, as Humboldt styles it, "a forest above a forest;" in every situation some are to be met with as representatives of the magnificent and regal family to which they belong.

In the following pages the genera and species are arranged in the order adopted by Dr. Martius in his elaborate work already alluded to.

NATURAL ORDER PALMACEÆ.
GENUS LEOPOLDINIA, *MARTIUS.*

This genus is characterized by having flowers containing stamens or pistils only, intermingled on the same spadix, and by not having a spathe. The male flowers have six stamens and no rudiments of a stigma. The female flowers have three sessile stigmas and rudimentary stamens. The spadix is much-branched and decomposed.

The species are trees of a moderate size without any spines or tubercles, but remarkable for the netted fibres which spring from the margins of the sheathing petioles, and cover the stem half-way down or sometimes even to its base. The leaves are terminal and pinnate, the leaflets spreading out regularly in one plane. There are often three or four spadices on a tree, bearing abundance of small flowers, and ovate compressed fruit, the outer covering of which is fleshy.

Four species are known, and they are all found in the same limited district near the Rio Negro, some extending to the tributaries of the Orinoco near its source, and one being found south of the Amazon nearly opposite the mouth of the Rio Negro. All however grow on the banks or in the immediate vicinity of black-water streams, which occur more extensively in South America than in any other part of the globe. Two species are described by Martius, one of which is here figured with two others, which are believed to be new. They are not found more than 1000 feet above the level of the sea.

LEOPOLDINIA PULCHRA. Ht. 12 Ft

PLATE IV.
LEOPOLDINIA PULCHRA, *MARTIUS.*
JARÁ, *LINGOA GERAL.*

THE Jará or Jará mirí (little Jará) is from ten to fifteen feet high. The stem is cylindrical, erect, and about two inches in diameter. The leaves are very regularly pinnate, about four feet long, with the leaflets slightly drooping and the terminal pair small. The leaf-stalks are slender and the sheathing bases are persistent, giving out from their margins abundance of flat fibrous processes which are curiously netted and interlaced together, clothing the stem with a firm covering often down to the very base. At the lower part this gradually rots and is rubbed away or falls off, leaving the stem bare. The flower-stalks or spadices are numerous, and very large and much branched; and the fruits are about an inch in diameter, oval and flattened, and of a pale green-ish-yellow colour. The outer covering is firm and fleshy, and has a very bitter taste.

This species is found on the banks of the Rio Negro and some of its tributaries, from its mouth up to its source, and on the black-water tributaries of the Orinoco. It never grows far from the water's edge, though generally out of reach of the floods in the wet season. It is not known to occur beyond this very limited district.

The stem of this tree being very smooth and cylindrical, and of a convenient length, it is much used for fencing round yards and gardens, and in the city of Barra do Rio Negro is universally employed for such purposes. The want of neatness out of doors, which is quite a characteristic of the Portuguese and Indian settlers on the Amazon, is always apparent in these fences. It is never thought worth while to cut the poles all to one length, but they are set up just as they are brought in from the forest; and the space between two handsome houses in the city may often be seen filled up with a Jará railing of most unpicturesque irregularity.

The bright green and glossy foliage of this tree also renders it suitable for another purpose. On certain saints' days, little altars and green avenues are made before the principal houses in Barra, the Jará palm being always used to construct them; and its graceful fronds rustling in the evening breeze, fitfully reflecting the light of the wax tapers which burn before the image of the saint, with the blazing torches of the rustic procession, have a very pleasing effect.

The reticulate covering of the stem of this and the next species offers a fine station for the epiphytal Orchideæ to attach themselves, and the Jará palms are accordingly often adorned with their curious and ornamental flowers.

Plate II. figure 6. represents a fruit of this species of the natural size.

30

LEOPOLDINIA MAJOR Ht 25 Ft

PLATE V.
LEOPOLDINIA MAJOR, N. SP.
JARÁ ASSÚ, *LINGOA GERAL.*

THE Jará assú or "greater Jará" closely resembles the last species, but it is considerably larger. The stem is four inches in diameter and reaches thirty feet in height. It is often much thicker at the bottom than in the upper part, and has a greater proportion of the stem bare. The leaves are very similar, but the spadices are larger, and the fruit is also larger and much more abundant.

This tree occurs plentifully on the lakes and inlets of the upper Rio Negro, but is not found at the mouth of the river like the last species. It grows too at a lower level, being often found with a part of the stem under water.

The Indians collect the fruit in large quantities, and by burning and washing extract a floury substance, which they use as a substitute for salt when they cannot procure that article. They assert positively that the smaller species of Jará will not yield the same product; but perhaps this may be only because the fruit is less abundant, and they do not take the trouble to collect it.

Coarse Portugal salt is used in the Rio Negro, and among the Indians in the upper part of the river serves as a circulating medium, about a pound of it being reckoned equivalent to a day's work. The supply however is very uncertain, and there are many distant tribes which it scarcely ever reaches; and it is among them that the substitute is manufactured from the fruit of the Jará. It is doubtful, however, whether it contains any true salt, for it is described as being more bitter than saline in taste; yet with this alone to season their fish and cassava the Indians enjoy almost perfect health. Perhaps, therefore, mineral salt may not be such a necessary of life as we are accustomed to consider it.

Pl. VI.

LEOPOLDINIA PIASSABA Ht 20 Ft

PLATE VI.
LEOPOLDINIA PIASSABA, N. SP.

Piassába, *Lingoa Geral.* Chíquichíqui, *Barré.* [An Indian language spoken on the Upper Rio Negro in Venezuela.]

THIS tree, the "Piassaba" of Brazil and the "Chíquichíqui" of Venezuela, I have little hesitation in referring to the genus *Leopoldinia,* though I have never seen it in flower or in fruit. The texture and form of the leaves, the peculiar branching of the spadix, and the extraordinary development of the fibres from the margins of the sheathing petioles, show it to be very closely allied to the other species of this genus.

The stem is generally short, but reaches twenty to thirty feet in height, and is much thicker than in either of the preceding species. The leaves are very large and regularly pinnate, with the pinnæ gradually smaller to the end, as in the two former species. The leaflets are rigid, broadest in the middle, and gradually tapering to a fine point, spreading out flat on each side of the midrib, but slightly drooping at the tips. The petioles are slender and smooth. The spadix is large, excessively branched and drooping, and there are often several on the same tree. The marginal processes of the petioles are interlaced as in the two former species, and are produced into long riband-like strips, which afterwards

35

split into fine fibres, and hang down five or six feet, entirely concealing the stem, and giving the tree a most curious and unique appearance. The leaves form an excellent thatch, and are almost universally used in that portion of Venezuela situated on the upper Rio Negro, and the adjacent tributaries of the Orinoco. The fruit is said to resemble that of the Jará in colour, but it is globose and eatable, being used principally to form a thick drink by washing off the outer coating of pulp.

The fibrous or hairy covering of the stem is an extensive article of commerce in the countries in which it grows. It seems to have been used by the Brazilians from a very early period to form cables for the canoes navigating the Amazon. It is well adapted for this purpose, as it is light (the cables made of it not sinking in water) and very durable. It twists readily and firmly into cordage from the fibres being rough-edged, and as it is very abundant, and is procured and manufactured by the Indians, piassaba ropes are much cheaper than any other kind of cordage. The price in the city of Barra in June 1852, was 400 reis or 1*s.* for 32 lbs. of the fibre, and 800 reis or 2*s.* for every inch in circumference of a cable sixty fathoms long, which is the standard length they are all made to.

Before the independence of Brazil, the Portuguese government had a factory at the mouth of the Paduarí, one of the tributaries of the Rio Negro, for the purpose of making these cables for the use of the Pará arsenal, and as a government monopoly. Till within these few years the fibre was all manufactured into cordage on the spot, but it is now taken down in long conical bundles for exportation from Pará to England, where it is generally used for street sweeping and house brooms, and will probably soon be applied to many other purposes. It is cut with knives by men, women and children, from the upper part of the younger trees, so as to secure the freshest fibres, the taller trees which have only the old and half-rotten portion within reach, being left untouched. It is said to grow again in five or six years, the fibres being produced at the bases of the new leaves. The trees are much infested by venomous snakes, a species of *Craspedocephalus,* and the Indians are not unfrequently bitten by them when at work, and sometimes with fatal consequences.

The distribution of this tree is very peculiar. It grows in swampy or partially flooded lands on the banks of black-water rivers. It is first found on the river Padauarí, a tributary of the Rio Negro on its northern side, about 400 miles above Barra, but whose waters are not so black as those of the Rio Negro. The Piassaba is found from near the mouth to more than a hundred miles up, where it ceases. On the banks of the Rio Negro itself not a tree is to be seen. The next river, the Darahá, also contains some. The next two, the Maravihá and Cababurís, are white-water rivers, and have no Piassaba.

On the S. bank, though all the rivers are black water, there is no Piassaba till we reach the Marié, not far below St. Gabriel. Here it is extensively cut for about a hundred miles up, but there is still none immediately at the mouth or on the banks of the Rio Negro. The next rivers, the Curicuríarí, the great river Uaupés, and the Isánna, though all black-water, have none; while further on, in the Xié, it again appears. On entering Venezuela it is found near the banks of the Rio Negro, and is abundant all up to its sources, and in the Témi and Atabápo, black-water tributaries of the Orinoco. This seems to be its northern limit, and I cannot hear of its again appearing in any part of the Amazon or Orinoco or their tributaries. It is thus entirely restricted to a district about 300 miles from N. to S. and an equal distance from E. to W. I am enabled so exactly to mark out its range, from having resided more than two years in various parts of the Rio Negro, among people whose principal occupation consisted in obtaining the fibrous covering of this tree, and from whom no locality for it can have remained undiscovered, assisted as they are by the Indians, whose home is the forest, and who are almost as well acquainted with its trackless depths as we are with the well-beaten roads of our own island.

The fibre imported into this country has been supposed to be produced only by the *Attalea funifera,* a species not found in the Amazon district. In the London Journal of Botany for 1849, Sir W. Hooker gave some account of the material, and of the tree producing it; stating that he had received the fruit of the tree with the fibre from a mercantile house connected with Brazil, and that the fruit was that of the *Attalea funifera*. This species is mentioned by Martius as furnishing a fibre used for cordage and other purposes in Southern Brazil, and he states that it is called "piaçaba"; so that the Indian name is applied to two distinct trees producing a similar material in different localities; and the two having been brought to England under the same name and from not very distant ports of the same country, were naturally supposed to be produced by the same tree. The greater part, if not all of the Piassaba now imported, comes, however, from the Rio Negro, where several hundred tons are cut annually and sent to Pará, from which place scarcely a vessel sails for England without its forming a part of her cargo.

Genus EUTERPE, *Gærtner.*

Male and female flowers intermingled on the same spadix, the former more abundant in the upper part of the branches, the latter in the lower. Spathe entire, membranaceous, fusiform and deciduous. Flowers with bracts, male with six stamens and a rudimentary pistil, female with three sessile stigmas. Spadix simply branched, spreading horizontally.

These are very elegant palms; their stems are lofty, slender, smooth and faintly ringed. The leaves are terminal, pinnate, regular, and form a graceful feathery plume. The bases of the petioles are sheathing for a long distance down the stem, forming a thick column three or four feet long, of a green or reddish colour. The spadices, three or four in number, spring from beneath the leaves, and the spathes are very deciduous, falling to the ground as soon as they open. The fruit is small, globose, at first green, then violet or black, and consists of a thin edible pulp covering the hard seed.

Twelve species are known, inhabiting the West Indies, Mexico and South America, and there appear to be three species in the Amazon district, two of which I have figured. Some prefer marshy grounds near the level of the sea, others extend up the mountains to a height of 4000 feet.

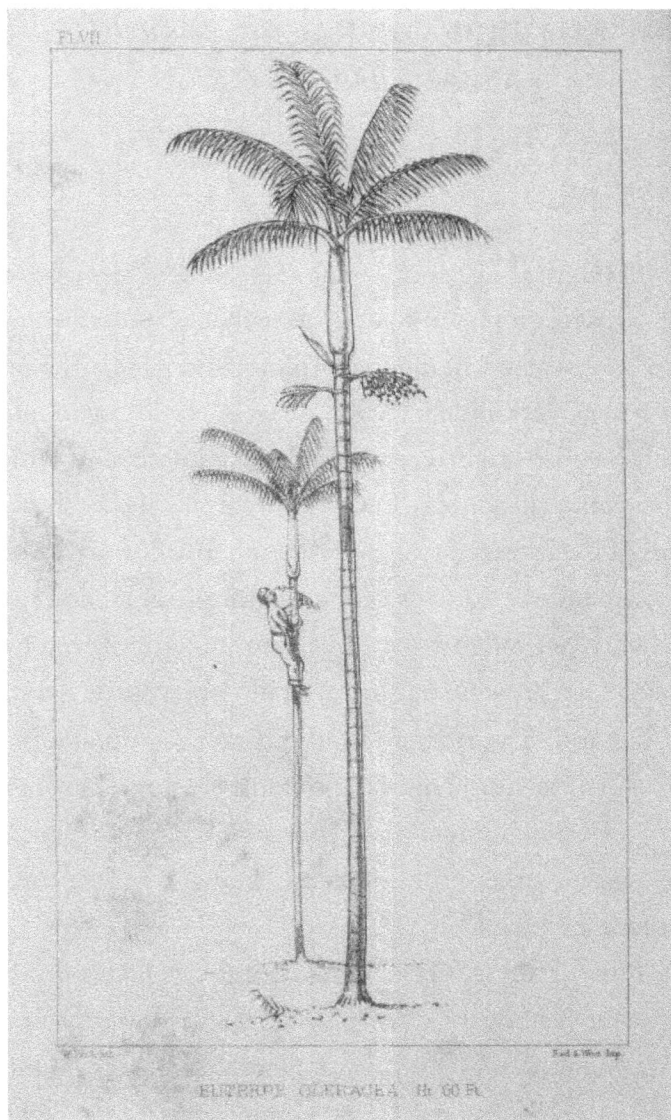

EUTERPE OLERACEA Ht 60 Ft

41

PLATE VII.
EUTERPE OLERACEA, *MARTIUS.*
ASSAÍ, *LINGOA GERAL.*

THE Assaí of Pará is a tall and slender tree, from sixty to eighty feet high, and about four inches in diameter. The stem is very smooth, of a pale colour, and generally waving, sometimes very much curved. The leaves are of moderate size, of a pale bright green, regularly pinnate, and with the leaflets much drooping. The column formed by the sheathing bases of the leaves is of an olive colour. The flowers are small, whitish, and very thickly set on the simply branched spadix. There are generally two or three, and sometimes even five or six spadices, growing out horizontally from a little below the leaf-column. The spathe is smooth and membranous, and falls off as the spadix opens. The fruit when ripe is about the size and colour of a sloe. It consists of a hard albuminous seed, with a rather fibrous exterior, and a very thin covering of a firm pulp or flesh.

This species is very abundant in the neighbourhood of Pará, and even in the city itself. It grows in swamps flooded by the high tides,—never on dry land. Its straight cylindrical stem is sometimes used for poles and rafters; but the tree is generally considered too valuable to be cut down for such purposes. A very favourite drink is made from the ripe fruit, and daily vended in the streets of Pará. Indian and negro girls may be constantly seen walking about with small earthen pots on their heads, uttering at intervals a shrill cry of Assaí—í. If you call one of these dusky maidens, she will set down her pot, and you will see it filled with a thick creamy liquid, of a fine plum colour. A penny-worth of this will fill a tumbler, and you may then add a little sugar to your taste, and will find a peculiar nut-flavoured liquid, which you may not perhaps think a great deal of at first; but, if you repeat your experience a few times, you will inevitably become so fond of it as to consider "Assaí" one of the greatest luxuries the place produces. It is generally taken with farinha, the substitute for bread prepared from the mandiocca root, and with or without sugar, according to the taste of the consumer.

During our walks in the suburbs of Pará we had frequently opportunities of seeing the preparation of this favourite beverage. Two or three large bunches of fruit are brought in from the forest. The women of the house seize upon them, shake and strip them into a large earthen vessel, and pour on them warm water, not too hot to bear the hand in. The water soon becomes tinged with purple, and in about an hour the outer pulp has become soft enough to rub off. The water is now most of it poured away, a little cold added, and a damsel, with no sleeves to turn up, plunges both hands into the vessel, and rubs and kneads with great perseverance, adding fresh water as it is required, till the whole of the purple covering has been rubbed off and the greenish stones left bare. The liquid is now poured through a wicker sieve into another vessel, and is then ready for use. The smiling hostess will then fill a calabash, and give you another with farinha to mix to your taste; and nothing will delight her more than your emptying your rustic basin and asking her to refill it.

The inhabitants of Pará are excessively attached to this beverage, and many never pass a day of their lives without it. They are particularly favoured too, in being able to get it at all seasons, for though in most places the trees only bear for a few months once in the year, yet in the neighbourhood of Pará there is so much variety of soil and aspect, that within a day or two's journey, there is always some ripe Assaí to supply the market. Boys climb up the trees to get it, with a cord round the ankles (as shown on the Plate), and with its own leaves make a neatly interlaced basket to carry it home. From the great island of Marajó, its igaripés* and marshes, from the rivers Guamá and Mojú, from the thousand islands in the river, and from the vast palm swamps in the depths of the forest, baskets of the fruit are brought every morning to the city, where half the population look to the Assaí to supply a daily meal, and hundreds are said to make it, with farinha, almost their main subsistence.

The trees of this genus also furnish another article of

food. The undeveloped leaves in the centre of the column form a white sweetish mass, which when boiled somewhat resembles artichoke or parsnep, and is a very good and wholesome vegetable. It may also be eaten raw, cut up and dressed as a salad with oil and vinegar. As, however, to obtain it the tree must be destroyed, it is not much used in Pará, except by travellers in the forest who have no particular interest in the preservation of the trees for fruit. The Cabbage Palm of the West Indies is an allied species, and is used for food in the same manner.

Very fine specimens of this tree may be seen in the great Palm House at Kew, where they grow almost as luxuriantly as in their native forests.

In the Plate, the unopened spathe, flower-spadix and fruit are represented, as they are often found, together on the same tree.

EUTERPE—?

On the banks of the Rio Negro there appears to be another species of this genus, closely allied to the *Euterpe oleracea,* but the stem is thicker and straighter, the whole tree larger, and the leaf-column thicker, and of a clear green colour. It grows on the dry land of the virgin forest, or sometimes within the limits of the winter's inundations. I unfortunately neglected to examine into its peculiar characters, as until my return to Pará I had considered it identical with the species so common there.

I was also informed that in the island of Marajó there is a species or variety having white fruit, but I had no opportunity of examining it.

Pl. VIII.

EUTERPE CATINGA Ht. 40 Ft.

47

PLATE VIII.
EUTERPE CATINGA, N. SP.
ASSAÍ DE CATINGA, *LINGOA GERAL.*

THIS species differs from the last in its slenderer stem and less drooping leaves and leaflets. It grows to forty or fifty feet high. The spadices are fewer and much smaller. The fruit also is smaller, and has more pulpy matter, so that a small quantity of it makes more of the "vinho d'Assaí" (the Assaí wine) than the same quantity of fruit of the larger kind. The column formed by the sheathing bases of the leaves is smaller than in the last species, and always of a red colour. The roots rise considerably above the ground, forming a distinct cone, which is not the case in the *E. oleracea.* It inhabits the forests on a dry sandy soil, of the Upper Rio Negro. These districts are called Catinga forests by the natives, and have very peculiar vegetable productions, differing almost entirely from those of the lofty virgin forest.

The preparation of the fruit of this species is sweeter and more finely flavoured than that of any other, and is therefore much sought after, but it takes the produce of four or five trees to yield as much as a single spadix of the larger kind will often produce. I found the fruit ripe in the month of April on the river Uaupés, a branch of the Rio Negro above the Falls.

GENUS ŒNOCARPUS, *MARTIUS.*

Male and female flowers on the same spadix, the former most abundant. Spathe double, the interior complete, woody, and deciduous. Flowers without distinct bracts; the male with six stamens and rudiments of a pistil, the female with three sessile stigmas, but with no rudiment of stamens.

These are tall majestic trees with large smooth stems, generally distinctly ringed. The leaves are large, terminal, more or less regularly pinnate, and have the bases expanded and clasping the stem, but not forming a sheathing column as in the last genus. The spadices spring from beneath the leaves and are simply branched; the branches are very lax, hanging down vertically except when forced outwards by the ripening fruit. The spathe is very large, fusiform and woody, and falls off the moment the spadix escapes from it. The fruit is small, nearly globular, and has an edible pulpy covering, like that of the genus *Euterpe.*

Six species only are known, and all inhabit tropical America, where they prefer dry, slightly elevated lands, none being known to extend more than 1600 feet above the sea.

Pl. IX.

OENOCARPUS BACCABA Ht 50 Ft

50

PLATE IX.
ŒNOCARPUS BACCÁBA, *MARTIUS.*
BACCÁBA, *LINGOA GERAL.*

THIS is a smooth thick-stemmed handsome tree, faintly ringed, and reaching fifty or sixty feet in height. The leaves are large, terminal, and pinnate. The leaflets are long; gradually pointed, and set at equal distances along the midrib. When young, the leaves are flat, the leaflets or pinnæ all standing out in the same plane; but in the full-grown tree the leaflets are in groups of two or three standing out at different angles from the general plane of the leaf, so as to give an irregular mixed appearance to the leaf. The petioles are greatly dilated at the base where they clasp the stem, and have a fibrous margin. The leaves as they die fall clean off from the stem, no part of the base remaining. The spathe is deciduous, being comparatively seldom visible. The fruits are of a violet or black colour when ripe, but are covered with a dense whitish bloom. They are prepared in the same way as the Assaí, but the pulp is of a pinkish cream-colour instead of purple, and the liquid is more oily, and of delicious flavour, somewhat resembling filberts and cream. It is said, however, not to be so wholesome as the Assaí, and in districts where intermittent fevers are prevalent, to bring them on, and to be particularly hurtful to persons recovering from that discase. A very beautiful oil is sometimes extracted from the pulp by pressure; it is perfectly clear, liquid, and inodorous; and serves as a substitute for olive oil, as well as being very good for lamps. The leaves are sometimes used for thatching when none better can be obtained; but owing to the irregularity of the pinnæ before mentioned, they are not much used.

This species inhabits the dry virgin forests of the Rio Negro and Upper Amazon. In the lower parts of that river and in the neighbourhood of Pará it is replaced by another species, the *œnocarpus distichus.*

The *œ. baccába* is growing at Kew.

One figure on the Plate shows the unopened spathe; the other has spadices with flowers and fruit.

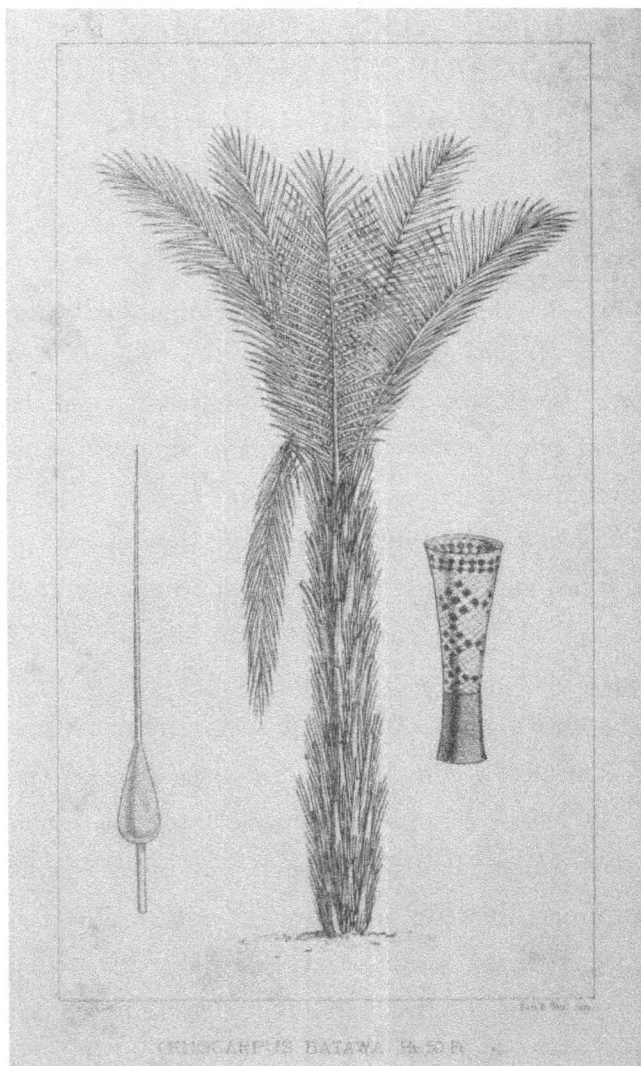

Pl. X.

OENOCARPUS BATAWA Ht 50 Ft

PLATES X. AND XI.
ŒNOCARPUS BATAWÁ, *MARTIUS.*
PATAWÁ, *LINGOA GERAL.*

THIS species can hardly be distinguished from the *œnocarpus baccába* when young. In the full-grown plant, however, the leaves preserve their regularity, the leaflets spreading out regularly in one plane and having a very beautiful appearance. The stem in old trees is fifty or sixty feet high and quite smooth, but in those growing in the shade of the forest, and in all young trees, the stem is completely hidden by the persistent bases of the decayed and fallen leaves. I have figured a tree in this state (Plate XI.).

The sheathing bases of the petioles give out from their margins numerous long spinous processes of a very singular character. They are from eighteen inches to three feet long, of a black colour, flattish, and generally broken or fibrous at the point. They are much sought after by the Indians, who use them to make arrows for their "gravatánas" or blow-pipes. One of these arrows is here represented with the wicker quiver in which they are carried. They are about fifteen or eighteen inches long, sharply pointed at the end, which is covered with "curarí" poison for three or four inches down, and notched so as to break off in the wound. Near the bottom a little of the soft down of the silk-cotton-tree is twisted round into a smooth spindle-shaped mass, and carefully secured with a fibre of a *"bromelia."* The cotton just fits easily into the tube, offering a light resisting body for the breath to act upon.

The fruit of this species is very similar to that of the Baccába, and is said to be of even superior flavour.

The Patawá is found in the whole of the Amazon and Rio Negro in the virgin forest, though apparently nowhere very abundant. Specimens are now growing in the Palm House at Kew.

The fruit is represented on Pl. X. of the natural size.

ŒNOCARPUS MINOR, *MARTIUS.*
BACCÁBA MIRI, *LINGOA GÉRAL.*

THIS is a small species common on the upper Rio Negro. The stem is not half so thick as in the *œ. baccába,* and the leaves are in proportion. The fruit is also very small, but is very fleshy and fine-flavoured, and ripens at a different time of year from the larger kind. It grows in the dry virgin forest. My drawing of this tree was unfortunately lost on my voyage home.

OENOCARPUS BATAWA Ht 60 Ft

Pl. XI.

OENOCARPUS BATAWA Ht 60 Ft

ŒNOCARPUS DISTICHUS, *MARTIUS*. BACCÁBA, OF PARÁ.

THIS is the species known as the Baccába at Pará, where the *œ. baccába* is not found. It is quite distinct from the allied species by the leaves being distichous, or arranged nearly in one plane on each side of the stem, which gives it a very peculiar aspect, unlike any other Palm.

On my return to Pará from the interior, I was suffering so much from ague, as to be unable to go in search of a specimen of this tree to figure as I had intended.

This, like all other species of the genus, grows in dry and rather elevated forest land.

GENUS IRIARTEA, RUIZ ET PAVON.

Female flowers few, interspersed among the males, bracteate. Spathe membranous, incomplete. Male flowers with from twelve to fifty stamens and the rudiments of a pistil. Female flowers with three sessile stigmas.

These singular and beautiful Palms have lofty, smooth,

cylindrical or ventricose stems, very faintly ringed. The roots grow more or less above ground. The leaves are terminal and pinnate, and the leaflets are somewhat triangular, notched, often twisted or curled, and have radiating nerves. The sheathing bases form a column as in *Euterpe*. The spadices grow from beneath the leaves and are simply branched and drooping. The spathes vary in number and size; they are membranous, and fall off before the fruit ripens. The fruit is oval, of moderate size, generally of a red or yellow colour, and the pulpy part is bitter and uneatable. The stems of this genus increase in thickness within certain limits, differing from most other palms, which, when the stem is once formed, only increase in height.

Nine species of this genus are known, all natives of South America. Four of them occur in the Amazon district, three in Bolivia, one in Venezuela, and one near Bogota, reaching a height above the sea of 5000 to 8000 feet.

Pl. XII.

IRIARTEA EXORHIZA. Ht. 60 Ft.

59

PLATE XII.
IRIARTEA EXORHIZA, *MARTIUS.*
PASHIÚBA, *LINGOA GERAL.*

This curious and beautiful tree is common in the forests about Pará and on the banks of the Amazon. It reaches fifty or sixty feet in height, with the stem moderately thick and very smooth, there being scarcely any rings or scars left by the fallen leaves.

The leaves are large and pinnate, with the leaflets triangular and very deeply notched, standing out at different angles with the midrib. The leaves curve over gracefully, and the character and aspect of the foliage is very different from that of most other palms. The column formed by the sheathing leaf-stalks is swollen at the base and of a deep green colour.

The spadices are three or four in number, growing rather upwards from the stem below the leaf-column. They are small and simply branched, and bear small oval red fruits about the size of a damson, the outer pulp of which is bitter and only eaten by some birds.

But what most strikes attention in this tree, and renders it so peculiar, is, that the roots are almost entirely above ground. They spring out from the stem, each one at a higher point than the last, and extend diagonally downwards till they approach the ground, when they often divide into many rootlets, each of which secures itself in the soil. As fresh ones spring out from the stem, those below become rotten and die off; and it is not an uncommon thing to see a lofty tree supported entirely by three or four roots, so that a person may walk erect beneath them, or stand with a tree seventy feet high growing immediately over his head.

In the forests where these trees grow, numbers of young plants of every age may be seen, all miniature copies of their parents, except that they seldom possess more than three legs, which gives them a strange and almost ludicrous appearance.

The figure on the opposite page (Plate XIII.) represents accurately the roots of a tree which had been partly blown down in the forest of the Upper Rio Negro. My friend Mr. Spruce informs me that it is a distinct species from that found at Pará, though closely allied to it, and scarcely differing in the character of the roots.

The wood of these trees is very hard on the outside, but soft and pithy within. It splits easily and very straight, and is much used for forming the floors of canoes, the ceilings of houses, shelves, seats, and various other purposes. Perfectly straight laths are more readily made from it than from any other wood, and they are so hard and durable as to serve for fish-weirs, corals for turtles, and for harpoons. The air-roots are covered with tubercular prickles, and are used by some Indians to grate their mandiocca.

This species grows in swamps or marshy ground in the virgin forest, not in the tide-flooded lands on the river banks.

Young plants may be seen in the great Palm House at Kew.

A fruit is represented on Plate III. fig. 5. of the natural size.

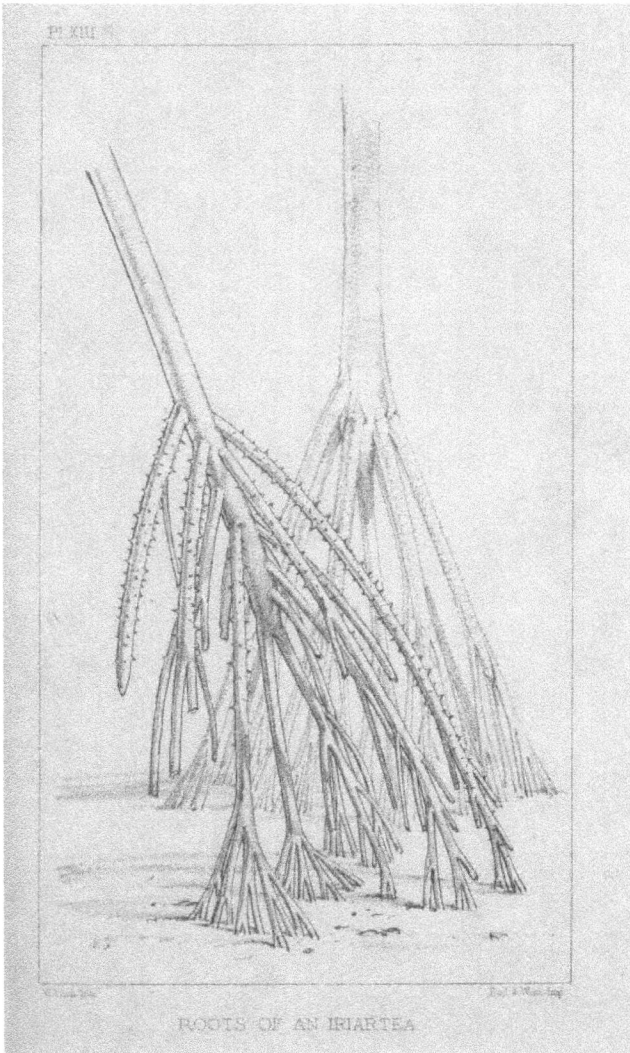

ROOTS OF AN IRIARTEA

Pl. XIII.

ROOTS OF AN IRIARTEA

IRIARTEA VENTRICOSA Ht 20 Ft

Pl. XIV.

IRIARTEA VENTRICOSA Ht 20 Ft

PLATE XIV.
IRIARTEA VENTRICOSA, *MARTIUS.*
PASHIÚBA BARRIGUDA, *BRAZIL.*

THIS is the most majestic tree of the genus. The stem reaches eighty or a hundred feet in height, and besides being rather thicker in proportion than in the last species, offers a remarkable character in being constantly more or less swollen near the middle or towards the top. The trunk is generally cylindrical to a height of forty or fifty feet, where it swells out to double its former diameter or more for ten or fifteen feet further, when it again diminishes and becomes cylindrical for about twenty feet to the summit. It is only when the trees have reached their full height or nearly so that the swelling commences. In a forest where they abound many may be seen of a large size, but quite cylindrical from top to bottom, while others present every degree of swelling from a just perceptible thickening to a most extraordinary enlargement. The column of air-roots in this species is six or eight feet high, forming a compact conical mass, the separate roots being more slender than in the *Iriartea exorhiza.*

The leaves are very large, with the leaflets broadly triangular and much cut and waved, forming a very elegant and yet massive head of foliage. The leaf-column is very thick, much swollen at the base, and of a deep bluish green colour.

The unopened spathes are lunate in shape and curved downwards, and the spadices are small and simply branched.

The wood of this tree is very hard, heavy and black, and is used by the Indians for making harpoons and spears with which they hunt the cow-fish. The swollen part of the stem is sometimes cut down and made into a canoe, when one is required in a hurry; otherwise it is not made use of.

The tree grows on the Upper Amazon and Rio Negro, on hill sides and on the banks of brooks and springs; and the Indians say that wherever it abounds salsaparilha will be found growing near.

A fruit is represented on the Plate of the natural size.

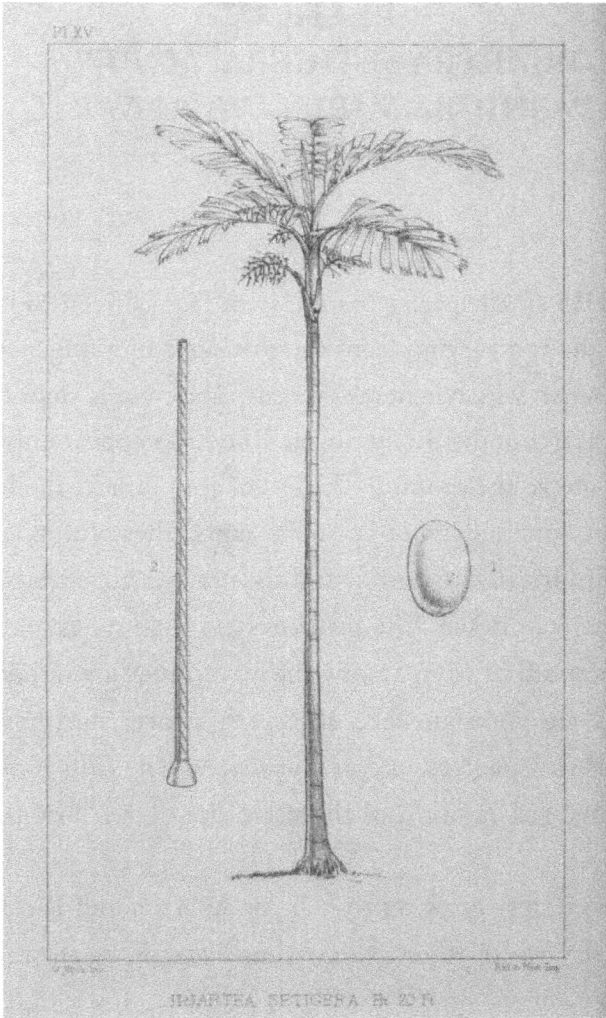

Pl. XV.

IRIARTEA SETIGERA Ht 20 Ft

PLATE XV.
IRIARTEA SETIGERA, *MARTIUS*.
PASHIÚBA MIRI, *LINGOA GERAL*.

THIS small species has the stem from fifteen to twenty feet high, and varying from the thickness of a finger to that of the wrist, which it never exceeds. The stem is smooth and cylindrical, but distinctly ringed. The roots appear only a few inches above the ground. The leaves are pinnate, the leaflets elongate, triangular and cut at the ends. The column is short and cylindrical, and both it and the petioles are covered with short hairs or down. The spadices have long stalks and grow from beneath or from among the leaves; they are rather large and are simply branched. The spathes form sheaths at the bases of the spadices, and are persistent. The fruit is oval, of an orange-red colour, and about the size of the "hip" or wild rose fruit.

These trees grow on the Upper Amazon and Rio Negro in the dry virgin forest, where they occur in small scattered groves.

This species is of great importance to the Indian of the Rio Negro. With its stem he constructs his "gravatána" or blowing tube, which, with the little arrows before described as made from the spines of the Patawá, forms a most valuable weapon, enabling him to bring down monkeys, parrots and curassow birds from their favourite stations on the summits of the loftiest trees of the forest.

When he wishes to make a "gravatána" he searches in the forest till he finds two straight and tall stems of the "Pashiúba miri" of such proportionate thicknesses that one could be contained within the other. When he returns home he takes a long slender rod which he has prepared on purpose, generally made of the hard and elastic wood of the "Pashiúba barriguda," and with it pushes out the pith from both the stems, and then with a little bunch of the roots of a tree fern, cleans and polishes the inside till the bore becomes as hard and as smooth as polished ebony. He then carefully inserts the slenderer tube within the larger, placing it so that any curve in the one may counteract that in the other. Should it still be not quite correct, he binds it carefully to a post in his house till it is perfectly straight and dry. He then fits a mouth-piece of wood to the smaller end of the tube, so that the arrow may go out freely at the other; and when he wishes to finish his work neatly, winds spirally round it from end to end, the shining bark of a creeper. Near the lower extremity he forms a sight with the large curved cutting tooth of the Paca *(Cœlogenus paca)*, which he fixes on with pitch, and the gravatána is then fit for use.

These tubes are never less than eight and are often ten or twelve feet long, and on looking through a good one, not the slightest irregularity can be detected from one end to the other. The bore is generally not large enough to admit the tip of the little finger, so that the breath more readily fills the whole tube and propels the arrow with great velocity. The vertical direction is that in which the surest aim can be taken, and for which the gravatána is best adapted. When birds are feeding at the top of a lofty tree where the result of a gun-shot would be doubtful, a skilful Indian will take his station beneath it, and with a puff from his powerful lungs, will send up his little poisoned arrows with unerring aim. The wounded birds sometimes turn giddy and drop in a few seconds, or fly away to a neighbouring tree and in a minute fall heavily to the ground, or try to pluck out the arrows with their beaks, which, however, invariably break in the wound. The hunter carefully marks the direction in which each one falls, and when his quiver is emptied of arrows or the tree of birds, walks round and gathers up the game. His weapon makes no noise, and he therefore often does more execution than the best European sportsman armed with his double-barrel Manton.

On Plate XV. fig. 1. is a fruit of the natural size; fig. 2. is the gravatána or Indian blowpipe.

GENUS RAPHIA, *COMMERSON.*

Male and female flowers intermixed on the same spadix. No common spathe, but many small incomplete sheaths. Male flowers with from six to twelve stamens and no rudiments of a pistil. Female flowers with three sessile stigmas and barren stamens.

The stems are short, thick and ringed. The leaves are very large, regular and pinnate; the leaflets are linear and have spinulose midribs and edges. The bases of the petioles are sheathing, and persistent some way down the stem, and the margins are fibrous. The spadices grow from among the leaves, and are very large and much branched; and the fruit is oblong and covered with large imbricated scales.

There are three species of the genus known; one is a native of the west coast of Africa, another of Madagascar, while a third is found on the banks of the Lower Amazon.

Pl. XVI.

RAPHIA TæDIGERA Ht 60 Ft

PLATE XVII.
RAPHIA TÆDIGERA, *MARTIUS.*
JUPATÍ, *LINGOA GERAL.*

THIS is one of the most striking of the many noble Palms which grow on the rich alluvium of the Amazon. Its comparatively short stem enables us fully to appreciate the enormous size of its leaves, which are at the same time equally remarkable for their elegant form. They rise nearly vertically from the stem and bend out on every side in graceful curves, forming a magnificent plume seventy feet in height and forty in diameter. I have cut down and measured leaves forty-eight and fifty feet long, but could never get at the largest. The leaflets spread out four feet on each side of the midrib. They are rather irregularly scattered and not very closely set; they droop at the tips and have weak spinules along the margins.

The stem does not generally exceed six or eight feet in height and is about a foot in diameter, clothed for some distance down with the persistent sheathing bases of the leaf-stalks and the numerous spinous processes which proceed from them. These spines are something like those of the "Patawá," but not so thick and strong.

The spadices are very large, compoundly branched

and drooping; they grow from among the leaves and have numerous bract-like sheaths in the place of spathes.

The flowers are of a greenish olive colour and densely crowded, and the fruit is large, oblong, and reticulated with large scales.

The petiole or leaf-stalk of this tree is most extensively useful. It is often twelve or fifteen feet long below the first leaflets, and four or five inches in diameter, perfectly straight and cylindrical. When dried, it almost equals the quill of a bird for strength and lightness, owing to its thin hard outer covering and soft internal pith. But it is too valuable to the Indian for him to use it entire. He splits off the smooth glossy rind in perfectly straight strips and makes baskets and window blinds. The remaining part is of a consistence between pith and wood, and is split up into laths about half an inch thick and serves for a variety of purposes. Window shutters, boxes, bird-cages, partitions and even entire houses are constructed of it. In the little village of Nazaré near Pará, many houses of this kind may be seen in which all the walls are of this material, supported by a few posts at the angles and fastened together with pegs and slender creepers (sipós).

The hand may be easily pushed through one of these walls, but as the inhabitants do not trouble themselves with the possession of any article worth stealing, they sleep as composedly as if stone walls and iron bolts shut them in with all the security of a more advanced civilization.

The same material is also used for stoppers for bottles, and we found it answer admirably for lining our insect boxes, holding the pins securely and being more uniform in its texture than cork.

This is the only American species of the genus, and it inhabits exclusively the tide-flooded lands of the Lower Amazon and Pará rivers, being quite unknown in the interior. When descending from the Rio Negro to Pará in the summer of 1852, I observed some of our Indians who had made the voyage before, pointing out this tree to their less travelled companions as one of the curiosities of the lower country not to be found in the "Sertaõ."

It is probable that the leaf, though not entire, is the largest in the whole vegetable kingdom, some of them covering a surface of more than 200 square feet. In a few years we may be able to see them in the magnificent Palm House at Kew, where young plants are now growing.

Plate II. fig. 1, a fruit of *Raphia tœdigera* of the natural size.

GENUS MAURITIA, *LINNŒUS.*

Male flowers on one tree, female or hermaphrodite flowers on another. The spathes are imperfect, bract-like, tubular sheaths. The male flowers have six stamens. The female flowers have a three-lobed stigma and six imperfect stamens.

The stems are either tall, columnar and smooth, or more slender and armed with strong conical spines. The leaves are all fan-shaped or radiating from a centre. The spadix is very large and pinnately branched, and grows from among the leaves. The fruits are of moderate size, oval or globular, and covered with rather small imbricated scales pointing downwards.

Four species are described by Martius, three of which occur in the Amazon district, and four more were met with by me on the Rio Negro, so that the genus seems confined to the hottest parts of the American Continent from the level of the sea to an altitude of about 3000 feet.

Pl. XVII.

MAURITIA FLEXUOSA. HT. 100 Ft

PLATE XVII.
MAURITIA FLEXUOSA, *LINNÆUS.*
MIRITÍ, *LINGOA GERAL.*
MURÍCHI, *IN VENEZUELA.*
ITÁ? *MOUTH OF THE ORINOCO.*

THIS is one of the most noble and majestic of the American Palms. It grows to a height of eighty or a hundred feet. The stem is straight and smooth, about five feet in circumference, often perfectly cylindrical, but sometimes swollen near the middle or towards the top, so that the bottom is the thinnest part.

The leaves spread out in every direction from the top of the stem. They are very large and fan-shaped, the leaflets spreading out rigidly on all sides and only drooping at the tips and at the midrib or elongation of the petiole. The leaves stand on long stalks which are very straight and thick, and much swollen at the base which clasps the stem. A full-grown fallen leaf of this tree is a grand sight. The expanded sheathing base is a foot in diameter; the petiole is a solid beam ten or twelve feet long, and the leaf itself is nine or ten in diameter. An entire leaf is a load for a man.

The spadices grow out from among the leaves; they are very large, pinnately branched and horizontal or drooping.

The fruit is spherical, the size of a small apple and covered with rather small, smooth, brown, reticulated scales, beneath which is a thin coating of pulp. A spadix loaded with fruit is of immense weight, often more than two men could carry between them.

The leaves, fruit and stem of this tree are all useful to the natives of the interior. The leaf-stalks are applied to the same purposes as those of the species last described, the Jupatí. The epidermis of the leaves furnishes the material of which the string for hammocks, and cordage for a variety of purposes is made. The unopened leaves form a thick-pointed column rising from the very centre of the crown of foliage. This is cut down, and by a little shaking the tender leaflets fall apart. Each one is then skilfully stripped of its outer covering, a thin riband-like pellicle of a pale yellow colour which shrivels up almost into a thread. These are then tied in bundles and dried, and are afterwards twisted by rolling on the breast or thigh into string, or with the fingers into thicker cords. The article most commonly made from it is the "réde," or netted hammock, which is the almost universal bed of the native tribes of the Amazon. These are formed by doubling the string over two rods or poles about six or seven feet apart, till there are forty or fifty parallel threads, which are then secured at intervals of about a foot by cross strings twisted and tied on to every longitudinal one. A strong cord is then passed through the loop formed by all the strings brought together at each end, by which the hammock is hung up a few feet from the ground, and in this open net the naked Indian.

sleeps beside his fire as comfortably as we do in our beds of down.

Other tribes twist the strings together in a complicated manner so that the hammock is more elastic, and the Brazilians have introduced a variety of improvements by using a kind of knitting needles producing a closer web, or by a large wooden frame with rollers, on which they weave in a rude manner with a woof and weft as in a regular loom. They also dye the string of many brilliant colours which they work in symmetrical patterns, making the rédes or "maqueiras" as they are there called, among the gayest articles of furniture to be seen in a Brazilian house on the Amazon.

From the fruits a favourite Indian beverage is produced. They are soaked in water till they begin to ferment, and the scales and pulpy matter soften and can be easily rubbed off in water. When strained through a sieve it is ready for use, and has a slight acid taste and a peculiar flavour of the fruit at first rather disgreeable to European palates.

In the tidal districts about Pará, the massive trunks of these trees are often used to form a raised pathway across the expanse of soft mud generally left at low water between "terra firma" and the water's edge. A smooth and slippery cylinder is certainly not the best thing that could be devised for this purpose, but as it is the most easily procured and the least expensive it is proportionately common, and on paying a visit to many a Brazilian country house, should you arrive at low water, you will have no other means of getting ashore.

The Miriti is a social palm, covering large tracts of tide-flooded lands on the Lower Amazon. In these places there is no underwood to break the view among interminable ranges of huge columnar stems rising undisturbed by branch or leaf to the height of eighty or a hundred feet,—a vast natural temple which does not yield in grandeur and sublimity to those of Palmyra or Athens.

Of the age of these noble trees we have no knowledge, but it is remarkable how uniform they appear in size, there often being not a single young tree over a considerable extent of ground, particularly in places now flooded daily by the tide. One would therefore imagine that the present trees sprung up when the ground was more elevated than at present, and that it has since gradually sunk (or the waters risen) till the conditions have become unfavourable for the growth of young plants, though not hurtful to those which had already attained a certain age. Whether such is the true explanation of the phænomenon can only be decided by continued observation on the spot.

Besides this species which is mentioned by Martius as

occurring at Pará, my friend Mr. Spruce ascertained that another closely allied palm, the *Mauritia vinifera,* also occurs there. On the Upper Amazon and Rio Negro a palm is found supposed to be the *M. flexuosa,* but it is not so lofty a tree, which may perhaps be accounted for by its growing on annually instead of diurnally flooded lands. It is believed to be the same species which Humboldt observed on the Serra Duida. The Itá palm growing on the delta of the Orinoco is also thought to be the same species. On the river Uaupes, a branch of the Upper Rio Negro, I observed an allied species called by the natives "Caraná assu." The stem was smooth and much more slender and waving, and the leaves much smaller.

Plants of the *Mauritia flexuosa* are growing in the Palm House at Kew.

On Plate XVII. a single leaf is represented, showing the flabellate form produced by abbreviation of the midrib.

Plate II. fig. 2. is a fruit of the natural size.

Pl. XVIII.

MAURITIA CARANA Ht 40 Ft

Palm Trees of the Amozon

PLATE XVIII.
MAURITIA CARANA, N. SP.
CARANÁ, *LINGOA GERAL.*

THIS is a large smooth-stemmed species allied to *M. flexuosa,* but quite distinct and hitherto undescribed. The stem is about a foot in diameter and from twenty to forty feet high, smooth and obscurely ringed. The leaves are very similar to those of the Mirití, but the leaflets are not so deeply divided, being united together at the base for one-third of their entire length, and much more drooping at the tips. The petioles are very large, straight and cylindrical; their dilated bases are persistent for a considerable distance down the stem, and their margins give out a quantity of fibres which clothe it as in the *Leopoldinia piassaba,* though rather less densely.

The spadices grow from among the leaves and are somewhat more erect and much smaller than in the Mirití, and the fruits are less abundant, smaller and slightly ovate.

The leaf-stalks of this species are used for the same purposes as those of the Mirití and Jupatí already described, as those palms are generally absent where this is abundant. The part most generally used, however, is the leaf, which for thatching is preferred to that of any other species, on

84

account of its having so large a portion of the base entire and being of a very durable texture. A roof well-thatched with Caraná will last eight or ten years without renewing, and the leaves are so constantly cut for this purpose that it is hardly possible to find an entire and handsome tree. Though so closely resembling the Mirití, the epidermis is never used for cordage, and on my asking an Indian the reason, he quite laughed at the idea, saying that it was quite impossible because the Caraná "did not produce any thread."

This tree grows in the district of the Rio Negro and Upper Orinoco, but is not found on the Amazon. It prefers the dry Catinga forests, or the sandy margins of streams out of reach of the highest floods. At Javita I observed it growing within a few yards of the Mirití, but still preserving all its distinctive characters.

It is called by the natives Caraná, the smaller prickly stemmed species being known by the name of Caranaí.

MAURITIA ACULEATA. Ht 45 Ft

Pl. XIX.

MAURITIA ACULEATA. Ht 45 Ft

86

PLATE XIX.
MAURITIA ACULEATA, *HUMBOLDT.*
CARANAÍ, *LINGOA GERAL (RIO NEGRO).*
CARANÁ? *(PARÁ).*

THIS species has a tall, erect and slender stem reaching about forty or fifty feet in height and armed with numerous, long, conical, woody spines arranged in rings. The leaves are rather small with the leaflets rigid and very slightly drooping at the tips, and united at the base for about one-eighth of their length. The petioles are long and slender and are deciduous, the entire leaf falling away from the stem. The midrib and edges of the leaflets are armed with weak spinules. The spadices are small and grow somewhat erect so as to be partly concealed among the leaves, and the fruit is oval and rather small.

This species grows on the Upper Rio Negro and Atabapo, in marshes, with a rocky subsoil, and in the moist parts of the Catinga forest. The Caraná, common in the swamps (not in the tide-flooded lands) about Pará, is very closely allied or may be the same species.

Pl. XX.

MAURITIA GRACILIS Ht. 30 Ft.

PLATE XX.
MAURITIA GRACILIS, N. SP.
CARANAÍ, *LINGOA GERAL.*

THIS very elegant species is rather smaller than the last. The stem is from twenty to thirty feet high, slender, waving, and ringed with conical spines rather smaller than in *M. aculeata.*

The leaves are from five to eight in number with much-drooping leaflets. The petioles are slender, short, and greatly dilated at the base. The spadices are three or four in number, growing from among the leaves, of very large size in proportion to the tree, much-branched and drooping. They bear great quantities of fruit, which is of an oval shape and nearly as large as that of the *Mauritia carana.*

This beautiful little palm is first met with about Barcellos on the Rio Negro, more than 300 miles up the river, and is thence common as far as the black-water tributaries of the Orinoco. It always grows close to the water's edge in clumps of thirty or forty individuals, and its drooping leaves of a pale hoary green colour, never so much crowded as to lose their distinct outline, with the bending clusters of rich brown fruit, render it one of the greatest ornaments of its native river. The fruit is eaten, after being softened by soaking some time in water.

It seems closely allied to *M. armata* of Martius, which is found much farther south, on the banks of the S. Francisco River, but is probably quite a distinct species.

MAURITIA PUMILA H.B.R.

Pl. XXI.

MAURITIA PUMILA. Ht. 10 Ft.

PLATE XXI.
MAURITIA PUMILA, N. SP.
CARANAÍ, *LINGOA GERAL.*

THIS curious little palm is only eight or ten feet high, and has the stem slender, ringed, and armed with strong conical spines. The leaves are rather small and few in number, and the leaflets are much shorter, broader and more rigid than in any other palm of this genus. The petioles are long and rather thick, much sheathing at the bases which are persistent, clothing the stem some distance down after the leaves have dropped away from them, a character not found in any other prickly-stemmed species. The spadix is very long, branched and drooping. The fruit was not seen.

I only met with this palm on the Upper Rio Negro in two localities on the sandy margins of rivers and lakes just above the limits of the winter floods.

GENUS LEPIDOCARYUM, *MARTIUS.*

Male flowers on one tree, female or hermaphrodite flowers on another. Spathes, imperfect, bract-like, tubular sheaths. The male flowers have six stamens. The female flowers have three sessile stigmas and six imperfect stamens.

The stems are very slender, unarmed with spines or tubercles and deeply ringed. The leaves are fan-shaped, and have slender petioles and long swollen sheaths. The spadices are elongate and pinnately branched, growing from among the leaves. The fruits are oblong and covered with imbricated scales.

These delicate and very rare little Palms scarcely differ botanically from the last genus. Two species only are known, inhabiting the dense virgin forests of the Upper Amazon and Rio Negro, where they appear to be very locally distributed.

Pl XXII.

LEPIDOCARYUM TENUE Ht 8 Ft.

Pl. XXII.

LEPIDOCARYUM TENUE. Ht. 8 Ft.

94

PLATE XXII.
LEPIDOCARYUM TENUE, *MARTIUS.*
CARANAÍ DO MATO, *OF THE RIO NEGRO.*

THIS, the smallest of the fan-leaved Palms, has a smooth, ringed, waving stem as thick as one's finger and six or eight feet high. Its dark green glossy leaves, with narrow drooping leaflets, grow on long and slender stalks which have their sheathing bases much swollen and lengthened.

The spadices are small and slender, and the fruits, which are not abundant, are scaled in the same manner as those of the Mauritias, and are about the size of a large hazel-nut.

This rare and elegant species grows in the gloomiest depths of the virgin forest of the Upper Rio Negro, generally at some distance inland from the rivers, and shaded by the loftiest forest trees.

Plate II. fig. 4. represents a fruit of this species of the natural size.

GENUS GEONOMA, *WILLDENOW.*

Male and female flowers on distinct trees, or rarely on distinct spadices of the same tree. Spathe small, incomplete.

Male flowers with six stamens and a rudimentary pistil. Female flowers with three stigmas and a six-toothed ring of abortive stamens.

These are small palms with slender, smooth, ringed, reed-like stems. The leaves are large, regularly or irregularly pinnate, with the leaflets broad, and the bases of the petioles sheathing. The spadices are slender and more or less branched, and the spathes are double but small and membranous. The fruits are small, round or ovate, and are not eatable.

There are thirty-three species of this genus known, all of small size, and inhabiting various parts of South America and Mexico, from the level of the sea to 2000 feet above it. Many species may be seen flourishing in the Palm House at Kew.

GEONOMA MULTIFLORA Ht 12 Ft.

Pl. XXIII.

GEONOMA MULTIFLORA Ht 12 Ft

97

PLATE XXIII.
GEONOMA MULTIFLORA, *MARTIUS.*
UBIMRÁNA, *LINGOA GERAL.*

THIS handsome species is from eight to fifteen feet high, and has the stem regularly ringed or jointed, giving it a reed-like appearance. The leaves are very large, regularly pinnate and gracefully drooping on every side. The leaflets are very regularly placed on the midrib, and the terminal pair are much larger and broader. The petioles are slender and smooth, and the sheathing bases have an expanded fibrous margin.

The spadices grow from among the lower leaves, and are short, erect and simply branched. The spathes are very small and concealed among the petioles. The fruit is small, ovate, and when ripe of a red colour.

This appears to be the *Geonoma multiflora* of Martius, but the species are so closely allied that without a comparison of specimens it is very difficult absolutely to identify them.

I have found it only in the Catinga forests of the Upper Rio Negro, where it occurs very sparingly.

A fruit is represented on the Plate of the natural size.

Pl. XXIV.

GEONOMA PANICULIGERA Ht 9 Ft

99

PLATE XXIV.
GEONOMA PANICULIGERA, *MARTIUS.*
UBIM DE COTIWIYA, *LINGOA GERAL.*

THIS is a species from six to nine feet high and very similar in appearance to the last. The leaves, however, have only three or four pairs of leaflets of irregular width, the terminal pair being always very large and broad, and the others not being always placed opposite each other on the midrib.

The spadix is large, much branched and somewhat drooping, and has a small, soft and inconspicuous basal spathe. The fruit is small and round.

This species grows in the same localities and in the same soil as the last, but is much more abundant. It appears to agree well with the *G. paniculigera* of Martius.

There is a very closely allied species abundant in certain parts of the flooded lands or "gapó" of the Rio Negro, which is much used for thatching. The leaves being cut, the leaf-stalks are doubled and hitched on side by side to a strip of "pashiúba," and secured with "sipós" (which are the air-roots of *Arums* and other plants). They are said to make one of the most durable kinds of roof, and are much used for covering the semicircular "toldas" of canoes. They are also considered the best material for lining baskets of salt, and persons often go several days' journey to procure them for both these purposes.

I had no opportunity of closely examining the species which produces these leaves, and which is called "Ubim," in contradistinction to the other allied species which are termed "Ubimrana" (false ubim), "Ubim de cotiwiya" (Agouti's ubim) and other such names, and all of which, though sometimes used as substitutes, are said to be much less durable.

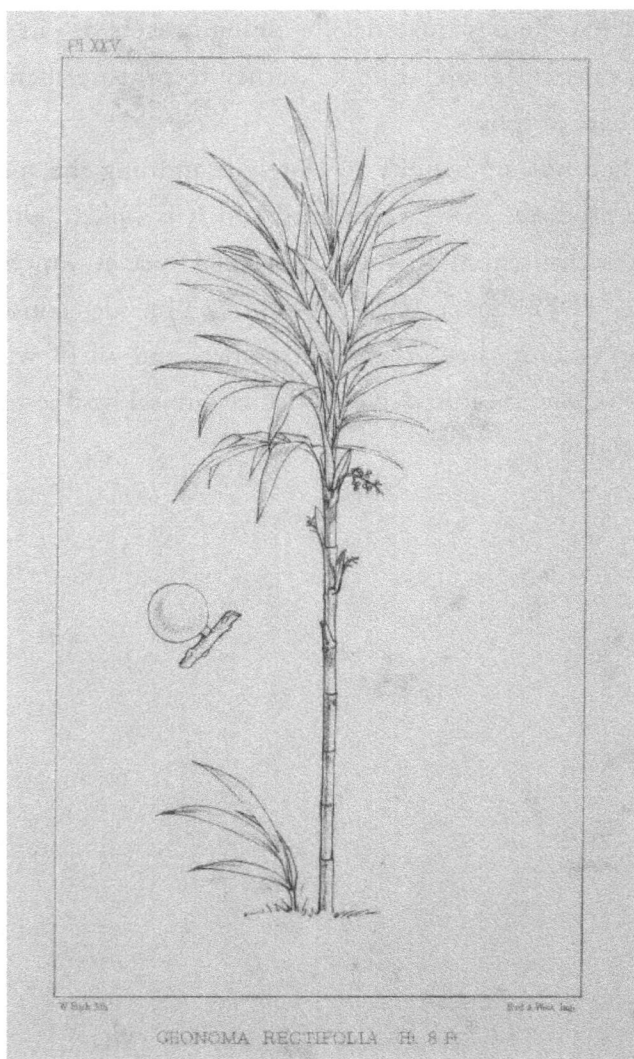

Pl XXV.

GEONOMA RECTIFOLIA Ht. 8 Ft

Pl. XXV.

GEONOMA RECTIFOLIA Ht. 8 Ft

PLATE XXV.
GEONOMA RECTIFOLIA, N. SP.
UBIMRÁNA, *LINGOA GERAL.*

THIS little species is nearly allied to the last. It reaches six or eight feet in height and has the stem distinctly jointed and the leaves persistent some way down it. The petioles grow very upright, and there are three or four pair of long, narrow and rather rigid leaflets, the terminal being the largest.

The spadices are numerous from the axils of the lower leaves, and are small and simply branched; and the fruit is very small, round and black.

This palm may be distinguished from *G. paniculigera,* to which it is most closely allied, by its very long narrow leaflets and much more erect habit; and by its smaller and less-branched spadices growing lower down on the stem, often below the leaves.

I found it in a few localities only on the Upper Rio Negro, growing in the sandy Catinga forest near the margin of the river.

A fruit is represented on the Plate of the natural size.

GENUS MANICARIA, *GŒRTNER.*

Male and female flowers in the same spadix. Spathe fusiform, fibrous, complete, breaking open irregularly. Male flowers with twenty-four to thirty stamens. Female flowers (situated below the male) with three sessile stigmas and twelve rudimentary stamens.

Stem short, thick and irregularly ringed. Leaves very large, entire and rigid, the sheathing bases persistent. Spadices simply branched, growing from among the leaves, nearly erect. Fruit large, hard, somewhat triangular or three-lobed and three-seeded, externally very rugose.

Only one species of this genus is known, which inhabits the Lower Amazon at the level of the sea.

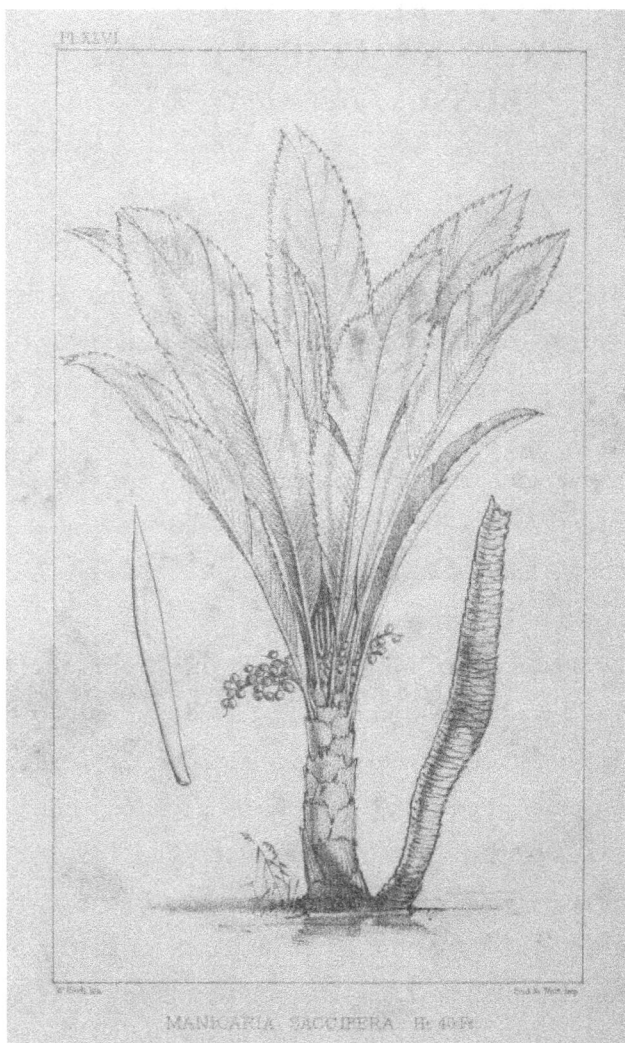

Pl. XXVI.

MANICARIA SACCIFERA Ht 40 Ft

105

PLATE XXVI.
MANICARIA SACCIFERA, *GŒRTNER.*
BUSSÚ, *LINGOA GERAL.*

THIS unique and handsome palm has the stem from ten to fifteen feet high, curved or crooked and deeply ringed. The leaves are very large, entire, rigid and furrowed, and have a serrated margin; they are often thirty feet long and four or five wide, and split irregularly with age. The petioles are slender with a broadly expanded fibrous-edged sheath at the base. These sheaths are persistent and often cover the stem down to the ground.

The spadices are numerous, growing from among the leaves, and are simply branched and drooping. The fruit is of an olive colour, somewhat three-lobed and with a rugose or papillate exterior covering. The spathe is fusiform and entire, of a fibrous cloth-like texture and of a brown colour. As the spadix expands it breaks open irregularly, but in some cases a dead unopened flower bunch is found enclosed in an entire half-rotten spathe, as if the vital powers of the plant had not been sufficient to tear asunder the tough fibrous sheath.

The "bussú" produces the largest entire leaves of any known palm, and for this reason, as well as on account of their firm and rigid texture, they form the very best and most durable thatch. The leaves are split down the midrib and the halves laid obliquely on the rafters, so that the furrows formed by the veins lie in a nearly vertical direction and serve as so many little gutters to carry off the water more rapidly. A well-made thatch of "bussú" will last ten or twelve years, and an Indian will often take a week's voyage in order to get a canoe-load of the leaves to cover his house.

The spathe too is much valued by the Indian, furnishing him with an excellent and durable cloth. Taken off entire it forms bags in which he keeps the red paint for his toilet or the silk cotton for his arrows, or he even stretches out the larger ones to make himself a cap,—cunningly woven by nature without seam or joining. When cut open longitudinally and pressed flat, it is used to preserve his delicate feather ornaments and gala dresses, which are kept in a chest of plaited palm leaves between layers of the smooth "bussu" cloth.

This species inhabits the tidal swamps of the Lower Amazon. A palm called "bussu" is also found on the Rio Negro and Upper Amazon, but it is of a smaller size and is probably a distinct species.

A spathe is represented on the Plate and a dead stem from which the leaves have entirely fallen.

Plate II. fig. 3, a fruit of *Manicaria saccifera* of the natural size.

GENUS DESMONCUS, *MARTIUS.*

Male flowers on the upper parts of the branches of the spadix, females on the lower. Spathe fusiform, woody, at length deciduous. Male flowers with six stamens and linear acute anthers. Female flowers with a short style and three stigmas and six small scaly rudiments of stamens.

Stems slender, flexible, climbing over shrubs or trees. Leaves alternate, pinnate, much sheathing, with long hooked spines in the place of the three or four terminal pair of leaflets. The spadices are axillary and simply branched, the spathes double, fusiform or ventricose, and the fruits are small, round, and generally red. The stems and leaves are more or less prickly.

Fourteen species of these curious Palms are found in various parts of South America, principally in the low lands, as they are not known at a greater height than 2000 feet above the level of the sea. They differ remarkably from all other American palms in their long climbing stems, in which they resemble the Calami or Canes so abundant in the East Indies.

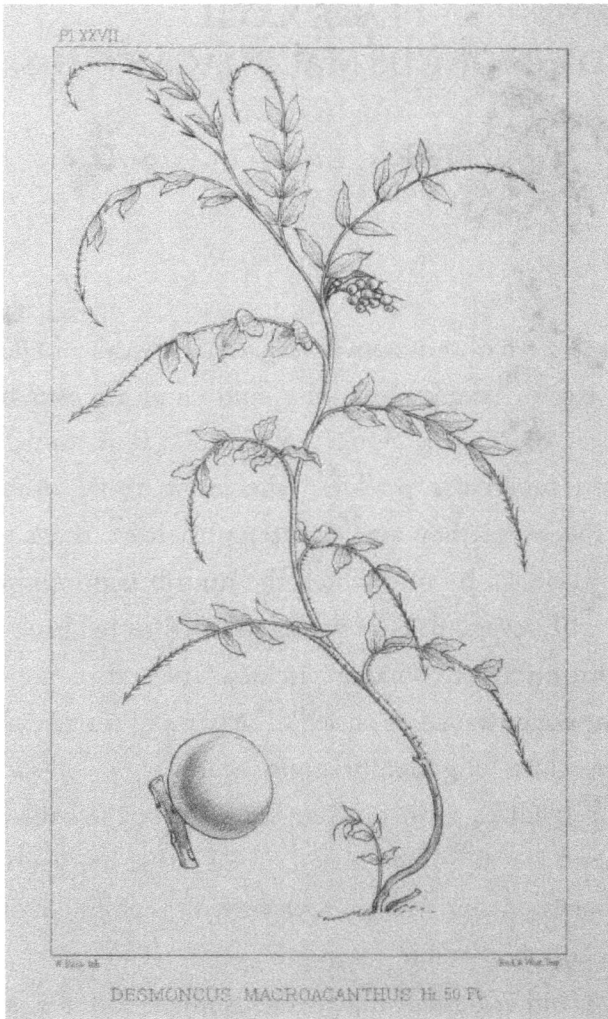

Pl XXVII.

DESMONCUS MACROACANTHUS Ht 50 Ft

Pl. XXVII.

DESMONCUS MACROCANTHUS Ht 50 Ft

PLATE XXVII.
DESMONCUS MACROACANTHUS,
MARTIUS.
JACITÁRA, *LINGOA GERAL.*

THE stem of this palm is very slender, weak and flexible, often sixty or seventy feet long, and climbing over bushes and trees or trailing along the ground. It is armed with scattered tubercular prickles. The leaves grow alternately along the stem; they are pinnate, with from three to five pairs of leaflets, beyond which the midrib is produced and armed with several pairs of strong spines directed backwards, and with numerous smaller prickles. The leaflets are ovate, with the edges waved or curled. The bases of the petioles are expanded into long membranous sheaths.

The spadices grow on long stalks from the axils of the leaves and are simply branched. The spathes are ventricose, erect, persistent and prickly, and the fruit is globular, of a red colour, and not eatable.

The rind or bark of this species is much used for making the "tipitis" or elastic plaited cylinders used for squeezing the juice out of the grated Mandiocca-root in the manufacture of farinha. These cylinders are some-times made of the rind of certain water plants and of the petioles of several palms, but those constructed of "Jacitára" are said to outlast two or three of the others, and though they are much more difficult to make, are most generally used among the Indian tribes. When the cylinders are used they are suspended from a strong pole, having been first filled with the grated pulp. A long lever is passed through the loop at the lower end of the "tipiti," by means of which it is stretched, the power being applied by a woman sitting on the further extremity of the pole. The cylinder thus becomes powerfully contracted, and the poisonous juice runs out at every part of the surface and is caught in a pan below in order to be carefully thrown away, for it would cause speedy death to any domestic animal which should drink it.

This species grows in the Catinga forests of the Upper Rio Negro and on the margins of small streams, climbing over trees and hanging in festoons between them, throwing out its armed leaves on every side to catch the unwary traveller. How often will they seize the insect-net of the ardent Entomologist just as he is making a dash at some rare butterfly, or fasten in his jacket or shirt-sleeve, or pull the cap from his head! Woe then to the impatient wanderer! a pull or a tug will inevitably cause a portion of the fractured garment to stay behind, for the "jacitára" never looses its hold, and it is only by deliberately extracting its fangs that the intruder can expect to depart unhurt.

In some places small igaripés or forest streams are almost

filled up with various climbing grasses and creepers, among which the "jacitára" holds a prominent place, and it is up these streams that the Indians often delight to fix their abode. In such cases they never cut down a branch, but pass and repass daily in their little canoes which wind like snakes among the tangled mass of thorny vegetation. They are thus almost safe against the incursions of the white traders, who often attack them in their most distant retreats, carry fire and sword into their peaceful houses and take captive their wives and children. But few white men can penetrate for miles along a little winding stream such as is here described, where not a broken twig or cut branch is found to show that a human being has ever passed before. Thus does the thorny "jacitára" help to secure the independence of the wild Indian in the depths of the forests which he loves.

This species most nearly agrees with the *D. macroacanthus* of Martius. Fine specimens of an allied species may be seen growing in the Palm House at Kew.

A fruit is represented on the Plate of the natural size.

GENUS BACTRIS, *JACQUIN.*

Male and female flowers intermingled in the same spadix, the females being more abundant in the lower parts and the males in the upper. Spathe double, exterior short and membranous, interior complete, woody. Male flowers with six, nine or twelve stamens. Female flowers with three sessile stigmas, and the stamens represented by a rudimentary ring.

The stems in this genus are very slender, ringed, nearly smooth or with a few scattered spines. The leaves are more or less terminal, generally few in number, pinnate or entire, with the bases of the petioles much sheathing and very spiny. The spathe is also clothed with spines. The spadices are simple or simply branched and grow from the axils of the leaves. The fruit is small and round, and the outer pulp is often subacid and eatable.

This very extensive genus of small prickly Palms contains forty-six species, all natives of South America. Two species described by Martius are here figured, together with six others apparently new, but as it may be impossible to identify those not seen in fruit, some of them have been left unnamed.

The species here figured are all from the Rio Negro, where I began studying them, and are sufficient to give an idea of their general characteristics and aspect.

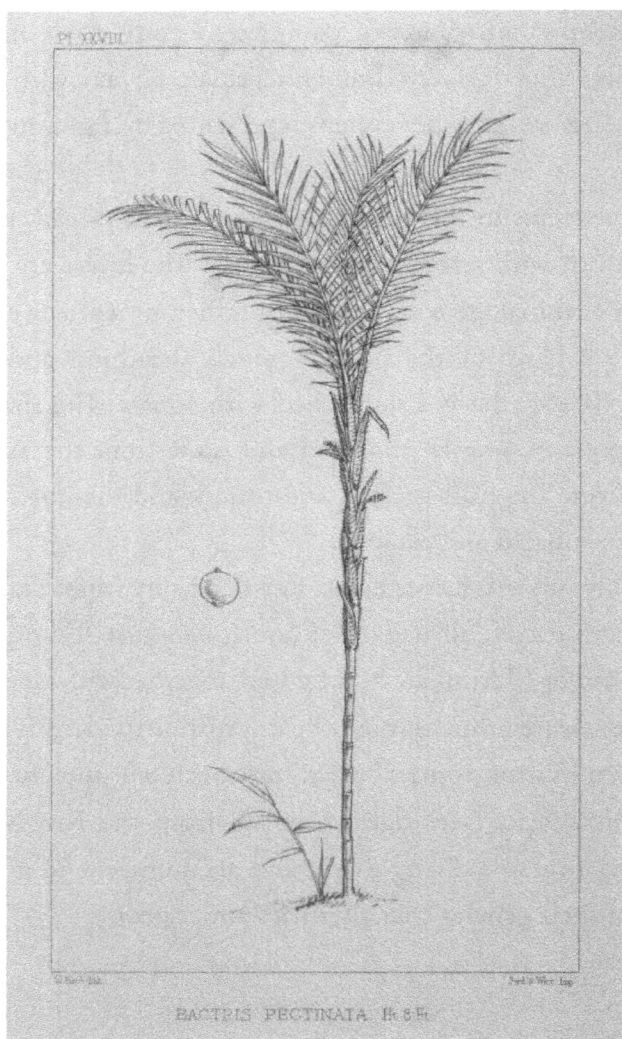

BACTRIS PECTINATA Ht 8 Ft

Pl. XXVIII.

BACTRIS PECTINATA Ht 8 Ft

PLATE XXVIII.
BACTRIS PECTINATA, *MARTIUS.*
IÚ, *LINGOA GERAL.*

THE stem of this species is from six to ten feet high, very slender, strongly ringed or jointed and smooth, but all other parts of the plant, the petioles, sheaths, spathes, &c., are prickly. The leaves are regularly pinnate, with the leaflets long, narrow, pointed and hairy beneath. The long sheathing bases of the petioles are persistent, covering the stem often half way down to the ground.

The spadices grow from among the persistent leaf-sheaths; they are very small, simple or two- or three-branched, and have a small persistent fibrous spathe. The fruit is very small and globular and of a red colour, and is not eatable.

This very hairy and prickly little palm grows in the sandy Catinga forest of the Upper Rio Negro and in the most exposed localities. It seems to agree well with the *B. pectinata* of Martius.

A fruit is shown on the Plate of the natural size.

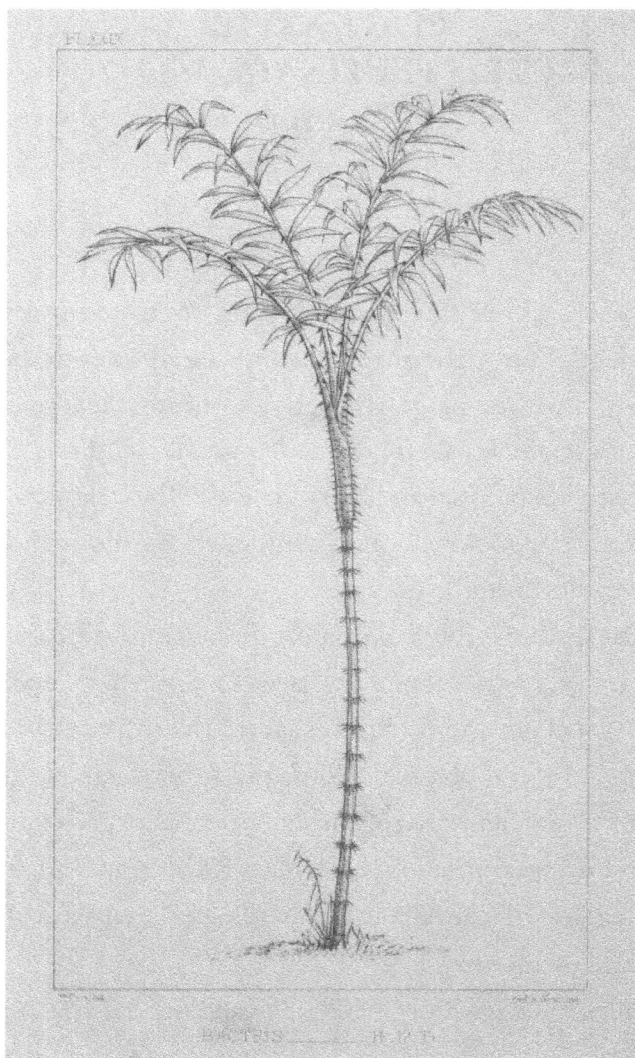

Pl. XXIX.

BACTRIS

116

PLATE XXIX.
BACTRIS—, N. SP.
MARAYARÁNA, *LINGOA GERAL.*

THE stem of this species is about an inch in diameter and ten or twelve feet high, thickly set with flat black spines disposed in rings. The leaves are rather large and irregularly pinnate, the leaflets being in little groups of two or four, standing out at various angles from the midrib, the groups themselves being set alternately along it. The leaflets are elongate and have the midrib produced in a bristly point, and the terminal pair are not larger than the rest. The petioles are armed with flat whitish spines, which on the long sheathing bases become black.

I met with this palm only once, growing in the dry virgin forest on the banks of the Rio Negro. Though it had neither flowers nor fruit at the time, yet its habit is so peculiar as to leave little doubt of its being a new species. It seems most nearly allied to the *Bactris macroacantha* of Martius.

Pl. XXX.

Fl.XXX

BACTRIS ELATIOR. Ht 20 Ft

Pl. XXX.

BACTRIS ELATIOR Ht 20 Ft

118

PLATE XXX.
BACTRIS ELATIOR, N. SP.
MARAYARÁNA, *LINGOA GERAL.*

THIS is a tall and elegant species. The stem is from fifteen to twenty feet high and about one inch in diameter, with a few scattered groups of small spines. The leaves are regularly pinnate, with broad leaflets narrowed at the base and ending in a lengthened point, the terminal pair being rather broader. The petioles and their sheathing bases are covered with broad, flat, whitish spines.

The spadices grow from among the lower leaves on long stalks and are simply branched and drooping. The spathes are elongate fusiform and spiny, and spiny, and are persistent. The fruit is small and globular.

This very graceful palm grows in the moist part of the virgin forest of the Upper Rio Negro, where I found it on the banks of small forest streams; and it seems quite distinct from any of the very numerous species described by Martius.

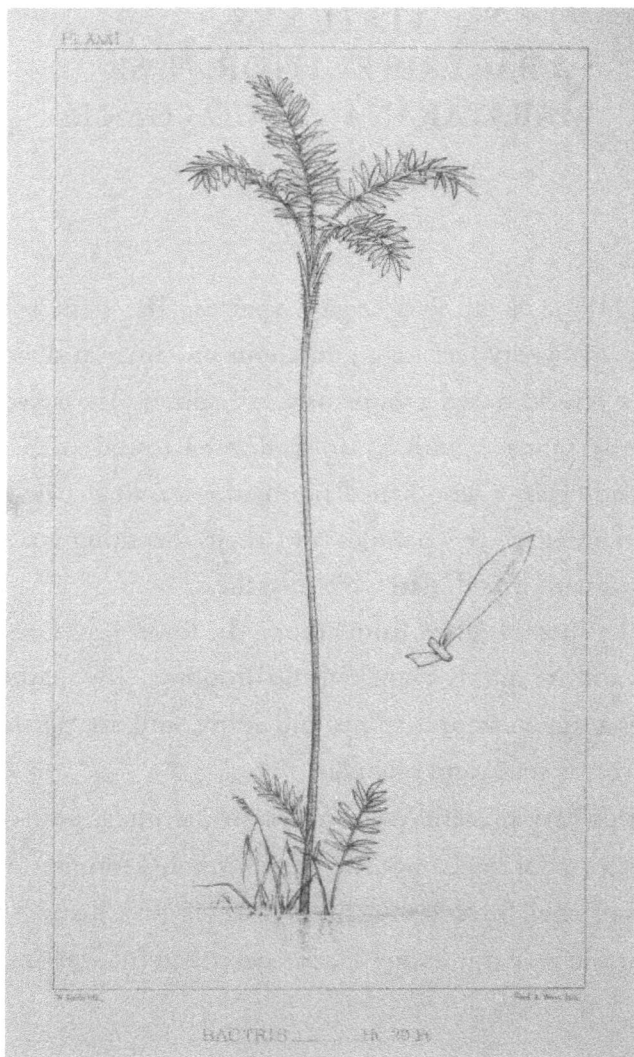

Pl. XXXI.

BACTRIS Ht 20 Ft

120

PLATE XXXI.
BACTRIS—, N. SP.
NATIVE NAME UNKNOWN.

THE stem of this curious palm is from twenty to twenty-five feet high and very slender. It is marked with slightly sunk rings and has a few scattered spines. The leaves are rather small, few in number and terminal. The leaflets are rigid, narrowed at the base, widest near the end and suddenly tapering to a point. They are arranged in groups of three or four at short intervals along the midrib, from which they stand out at different angles. The petioles and their sheathing bases are thickly set with slender, flattish, black spines generally pointing downwards.

This species was only found once, growing in the "gapó" or flooded lands of the Upper Rio Negro, and at the time had neither flowers nor fruit. The form and arrangement of the leaflets are so peculiar that it cannot be confounded with any species yet described.

A leaflet is represented of a larger size to show the peculiar form.

Pl. XXXII.

BACTRIS MACROCARPA Ht 10 Ft

122

PLATE XXXII.
BACTRIS MACROCARPA, N. SP.
IÚ, *LINGOA GERAL.*

THIS species has the stem about an inch in diameter and ten or twelve feet high, distinctly jointed, smooth and reed-like, but with a few spines in small groups at the joints. The leaves are terminal, of moderate size and rather interruptedly pinnate. The leaflets often grow in pairs and are broad, narrowed at the base and have the midrib produced at the point, the terminal pair being the largest. The petioles and sheaths are thickly set with whitish flat prickles.

The spadices are small, five- or six- branched, and rather long-stalked. The spathe is small, smooth and persistent. The fruit is oval, with a produced apex, large in proportion to the tree, of a reddish or yellowish olive colour, and not eatable, the outer covering being dry and woolly.

The smooth reed-like stem of this species resembles those of the Geonomas, and it is also remarkable for the large size of its fruit. It grows on the dry sandy soil of the Catinga forests of the Upper Rio Negro. It seems most nearly allied to *B. mitis* of Martius.

A fruit is represented on the Plate of the natural size, and a leaflet reduced one-fourth to show the peculiar form.

123

BACTRIS TENUIS Ht 6 Ft

Pl. XXXIII.

BACTRIS TENUIS Ht 6 Ft.

PLATE XXXIII.
BACTRIS TENUIS, N. SP.
IÚ, *LINGOA GERAL.*

IN this species the stem is not thicker than a goose quill, distinctly jointed and smooth. The leaves are terminal, four or five in number, and rather irregularly pinnate. The leaflets are elongate and acute, with produced points, four or five in number, on each side of the midrib, the terminal pair being the broadest. The petioles and their sheathing bases are covered with small, flat, black spines.

The spadices grow from below the leaves and are very small and unbranched. The spathes are fusiform, erect, persistent and smooth. The fruit is small, globose, and of a red colour.

This is one of the smallest of Palms, and in every part of its structure offers a striking contrast to the great *Mauritia* and other giants of the family. While they possess huge columnar stems a hundred feet in height and two feet in diameter, this has but a slender stalk the thickness of a quill; and while their fruit bunches are the largest in the vegetable kingdom, the whole spadix of this species is smaller than a bunch of currants.

It is allied to *B. cuspidata* and to *B. fissifrons* of Martius, but seems sufficiently distinct from either of them. It grows exposed to the sun in the sandy Catinga forests of the Upper Rio Negro.

An entire spadix with fruit is represented on the Plate, of the natural size.

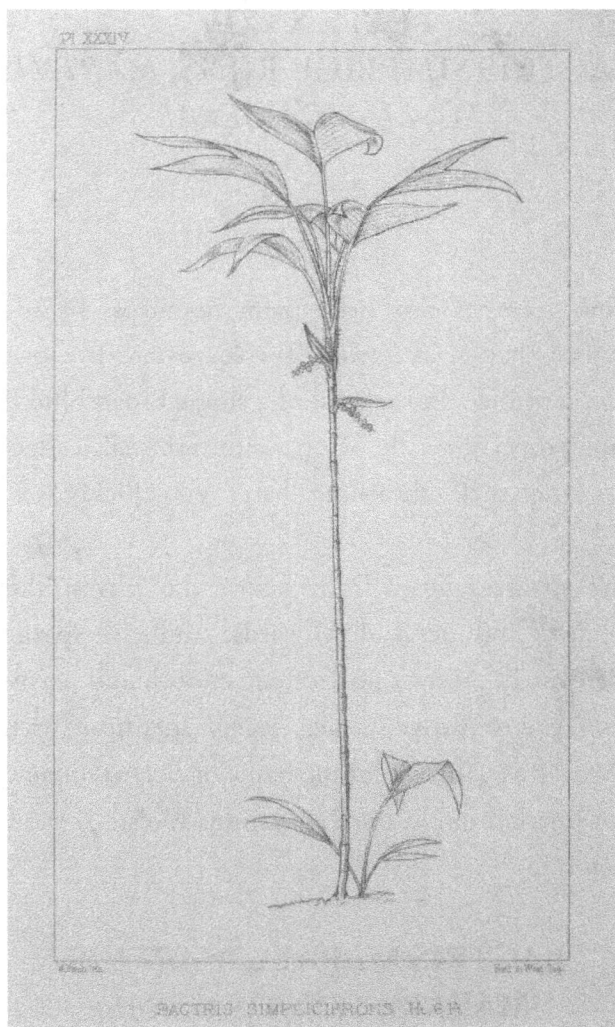

Pl. XXXIV.

BACTRIS SIMPLICIFRONS. Ht. 6 Ft.

127

PLATE XXXIV.
BACTRIS SIMPLICIFRONS, *MARTIUS.*
IÚ, *LINGOA GERAL.*

THE stem of this little palm resembles in size and appearance that of *B. tenuis.* The leaves are five or six in number, terminal, and consist of a single broad bifid leaflet, or more properly a pair of opposite terminal leaflets. The petioles and their sheathing bases are thickly set with spines.

The spadices grow from below the leaves; they are unbranched and bend downwards, and the spathes are elongate, small, erect or horizontal, smooth and persistent.

This pretty little species seems identical with one described by Martius under the name of *Bactris simplicifrons.* It is not uncommon in the dry Catinga forests of the Upper Rio Negro.

BACTRIS MARAJA, *MARTIUS.*
MARAJÁ, *LINGOA GERAL.*

This is a palm rather larger than most others of the genus, and inhabiting the flooded banks of the Amazon.

It produces large clusters of fruit resembling small black grapes, and having a thin pulp of an agreeable subacid flavour,—a peculiarity not found in the fruit of any other American palm that I am acquainted with. The places where it grows are often so deeply flooded that the fruit hangs close to the surface of the water, and can be plucked while passing in a canoe.

Dried specimens of the tree and fruit are in the Museum, and young plants are growing in the Palm House at Kew.

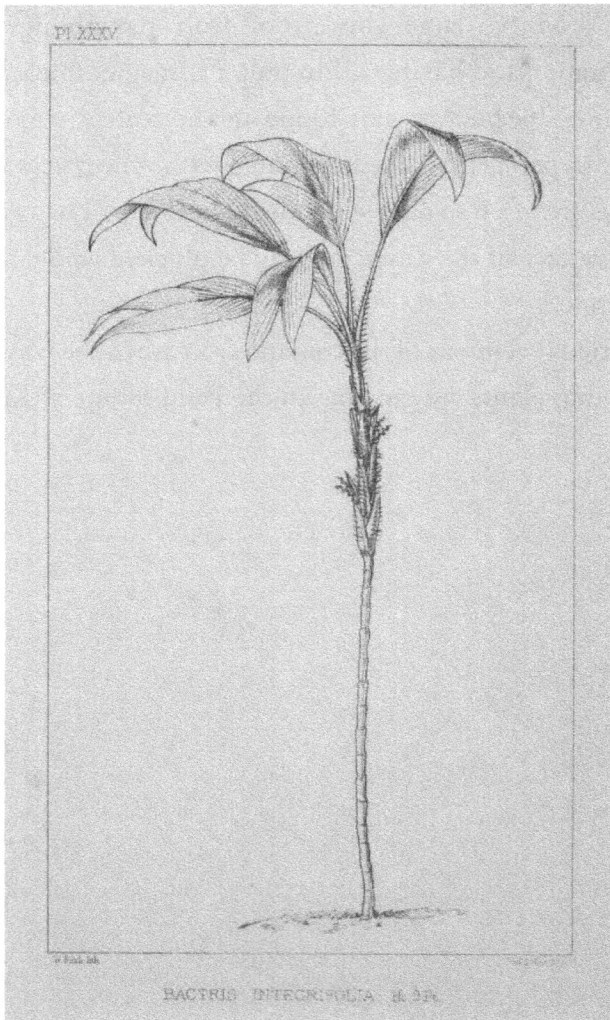

Pl XXXV

BACTRIS INTECRIFOLIA H 9Ft.

Pl. XXXV.

BACTRIS INTECRIFOLIA, Ht 9Ft.

PLATE XXXV.
BACTRIS INTEGRIFOLIA, N. SP.
IÚ, *LINGOA GERAL.*

THIS beautiful species has the stem hardly so thick as the little finger, and nine or ten feet high, smooth and distinctly jointed. The leaves are four or five in number, terminal, entire, three or four times as long as they are wide, and not very deeply bifid at the end. The petioles and their sheathing bases are thickly set with long, flat, black spines.

The spadices are very small, erect and two-branched, growing from among the persistent sheathing bases below the leaves. The spathes are small, erect and persistent, clothed with adpressed brown spines. The fruit is small and globular, and of a black colour.

This palm was found at S. Carlos on the Upper Rio Negro and on the "Estrada de Javita," a road through the forest for ten miles, which connects the river-systems of the Rio Negro and Orinoco, and along which most of the traffic between Venezuela and Brazil passes. In both cases it grew in the shady virgin forest.

GENUS GUILIELMA, *MARTIUS.*

Male and female flowers mixed in the same spadix, bracteate. Spathe double; exterior bifid; interior complete, woody. Male flowers with six stamens and a rudimentary pistil. Female flowers with three sessile stigmas, but with no rudiments of stamens.

The stems are lofty, rather slender, and armed with dense black cylindrical spines disposed in regular rings. The leaves are terminal and pinnate, but in the young plants entire, and the petioles are very spiny. The spadices are simply branched, growing from beneath the leaves, and the fruits are large, ovate, fleshy or mealy and eatable.

Three species only of this genus are known, inhabiting the lower mountain ranges of Peru and New Granada. They are lofty and conspicuous Palms with a remarkably handsome crown of foliage. One species only is found in the Amazon district, in all parts of which it is commonly cultivated.

Pl. XXXVI.

GUILMA . Ht. 60 Ft.

PLATE XXXVI.
GUILIELMA SPECIOSA, *MARTIUS.*
PUPÚNHA, *LINGOA GERAL.*
PIRIJAO, *INDIANS OF VENEZUELA,*
HUMBOLDT.
"THE PEACH PALM."

THIS most picturesque and elegant palm has the stem slender, cylindrical, and thickly set with long needleshaped spines disposed in rings or bands. It reaches sixty feet in height, and grows quite erect, though in exposed situations it becomes curved and waving. The leaves are very numerous, terminal, pinnate and drooping, forming a nearly spherical crown to the stem; and the leaflets growing out from the midrib in various directions, and being themselves curled or waved, give the whole mass of foliage a singularly plumy appearance. The young plants have the leaves entire like those of the Bussú, but as the age of the tree increases they break up into regular narrow leaflets.

The spadices grow from beneath the leaves, and are small, simply branched and drooping. The spathes are ventricose, woody and persistent, curving over the spadix.

The fruit is about the size of an apricot, of a triangular oval shape, and fine reddish-yellow colour. In most instances the seed is abortive, the whole fruit being a farinaceous mass. Occasionally, however, fruits are found containing the perfect stony seed, and they are then nearly double the usual size. This production of undeveloped fruits may be partly due to change of soil and climate, for the tree is not found wild in the Amazon district, but is invariably planted near the Indians' houses. In their villages many hundreds of these trees may often be seen, adding to the beauty of the landscape, and supplying the inhabitants with an abundance of wholesome food. In fact it here takes the place of the cocoa-nut in the East, and is almost as much esteemed.

As the stems are so spiny, it is impossible to climb up them to procure the fruit in the ordinary way. The Indians therefore construct rough stages up the sides of the trees, or form rude ladders by securing cross pieces between two of them, by which they mount so high as to be able to pull down the bunches of fruit with hooked poles.

The fruits are eaten either boiled or roasted, when they somewhat resemble Spanish chestnuts, but have a peculiar oily flavour. They are also ground up into a kind of flour, and made into cakes which are roasted like cassava bread; or the meal is fermented in water and forms a subacid creamy liquid. Parrots, macaws and many other fruit-eating birds devour them, and tame monkeys eat them greedily, though the wild ones cannot climb the spiny stems to obtain them.

The wood of this tree when old and black is exceedingly

hard, turning the edge of any ordinary axe. When descending the River Uaupes in April 1852, I had a number of parrots whose objections to any restraint upon their liberty caused me much trouble. Their first cage was of wicker, and in a couple of hours they had all set themselves at liberty. Then tough green wood was tried, but the same time only was required to gnaw that through. Thick bars of deal were bitten through in a single night, so I then tried the hard wood of the Pashiúba. This checked them for a short time, but in less than a week by continual gnawing they had chipped these away and again escaped. I now began to despair; no iron for bars was to be procured and my resources were exhausted, when one of my Indians recommended me to try Pupúnha, assuring me that if their beaks were of iron they could not bite that. A tree was accordingly cut down and bars made from it, and I had the satisfaction of seeing that their most persevering efforts now made little impression.

The very sharp needle-like spines of this tree are used by some tribes to puncture the skin, in order to produce the tattooed marks with which they decorate various parts of their bodies. Soot produced from burning pitch rubbed into the wounds is said to make the indelible bluish stain which these markings present.

This palm appears to be indigenous to the countries near the Andes. On the Amazon and Rio Negro it is never found wild. It is mentioned by Humboldt as having a smooth polished stem, which is a mistake.

Very fine specimens of this tree are growing in the great Palm House at Kew.

Plate III. fig. 4. represents a fruit of the natural size.

GENUS ACROCOMIA, *MARTIUS.*

Female flowers in the inner, male flowers in the outer part of the same spadix. Spathe complete, woody. Male flowers with six stamens and a rudimentary pistil. Female flowers with a short style and three stigmas, and a ring of abortive stamens.

The stems of these Palms are tall, strong, and more or less prickly. The leaves are large, pinnate, much drooping, and forming a dense spherical head of foliage. The leaflets are linear, and with the petioles are very prickly. The spadix is simply branched, and the fruit is round or oval, of an olive-green colour, and has a firm fleshy outer covering, which is often eaten.

Eight species of this genus are known, inhabiting various parts of South America, but more particularly Brazil. One or perhaps two species are found at Pará, but none on the Upper Amazon, where the alluvial soil and dense forests are unsuited to their growth.

ACROCOMIA LASIOSPATHA No 16 14

Pl. XXXVII.

ACROCOMIA LASIOSPATHA. Ht 40 Ft

PLATE XXXVII.
ACROCOMIA LASIOSPATHA, *MARTIUS.*
MUCUJÁ, *LINGOA GERAL.*

THE stem of this tree is about forty feet high, strong, smooth and ringed. The leaves are rather large, terminal and drooping. The leaflets are long and narrow, and spread irregularly from the midrib, every part of which is very spiny. The sheathing bases of the leaf-stalks are persistent on the upper part of the stem, and in young trees clothe it down to the ground.

The spadices grow from among the leaves, erect or somewhat drooping, and are simply branched. The spathes are woody, persistent and clothed with spines. The fruit is the size of an apricot, globular, and of a greenish-olive colour, and has a thin layer of firm edible pulp of an orange colour covering the seed.

This species is common in the neighbourhood of Pará, where its nearly globular crown of drooping feathery leaves is very ornamental. The fruit, though oily and bitter, is very much esteemed and is eagerly sought after. It grows on dry soil about Pará and the Lower Amazon, but it is quite unknown in the interior.

Several young plants of this and a species closely

resembling it, the *A. sclerocarpa,* are growing in the Palm House at Kew, and in the Museum at the same place are specimens of the stem and fruit sent by Mr. Bates and myself from Pará.

Martius mentions the *A. sclerocarpa* only as being found at Pará, but his description of the other species agrees best with the tree here figured. The two, however, seem very closely allied, if they are really distinct species.

A fruit is represented on the Plate of the natural size.

GENUS ASTROCARYUM, *MEYER.*

Female flowers few in number, situated beneath the males on the same spadix. Spathe complete, woody. Male flowers with six stamens and a rudimentary pistil. Female flowers with three stigmas and a rudimentary ring of stamens.

In this genus the stems are generally lofty and thickly set with rings of spines, but some species are stemless. The leaves are large and pinnate, the leaflets elongate and linear, and as well as the petioles very prickly. The spadices are simply branched, and the fruits are ovate or globose, with a fibrous or fleshy covering, sometimes eatable.

Sixteen species of these Palms are known, inhabiting Mexico, Brazil, and other parts of South America, but not extending higher up the mountains than 2000 feet above the sea. They have rather a repulsive aspect, from almost every part,—stem, leaves, fruit-stalk and spathe, being armed with acute spines in some cases a foot long.

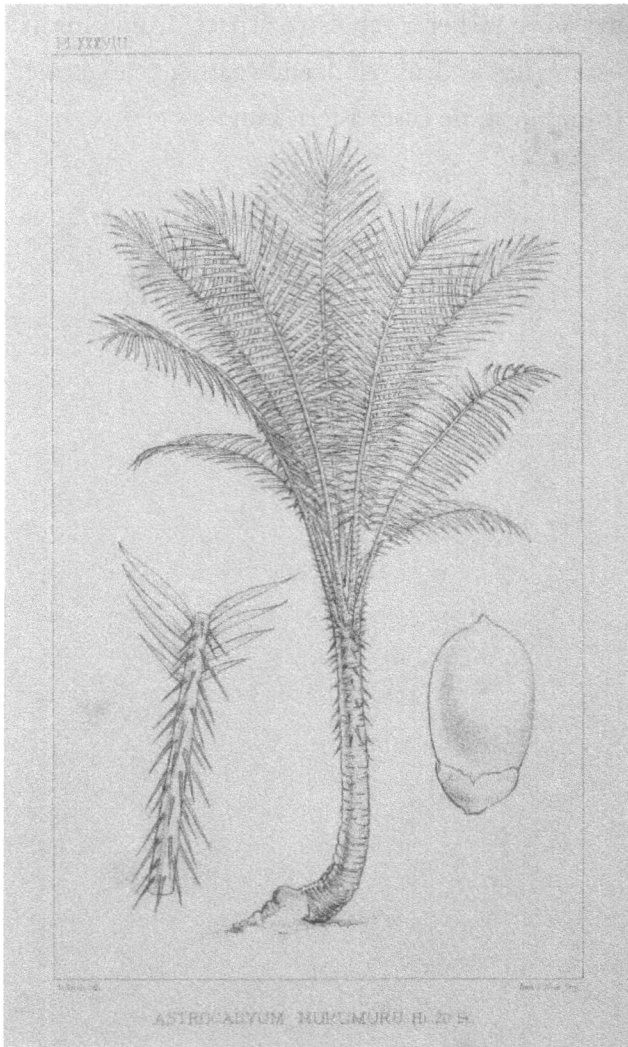

Pl. XXXVIII.

ASTROCARYUM MURUMURU Ht 20 Ft

142

PLATE XXXVIII.
ASTROCARYUM MURUMURÚ, *MARTIUS.*
MURUMURÚ, *LINGOA GERAL.*

THIS palm has the stem from eight to twelve feet high, irregularly ringed, and armed with long scattered black spines. The leaves are terminal and of moderate size, regularly pinnate, the leaflets spreading out uniformly in one plane, elongate, acute, with the terminal pair shorter and broader. The petioles and sheathing bases are thickly covered with long black spines generally directed downwards, and often eight inches long.

The spadices grow from among the leaves and are simply branched and spiny, erect when in flower, but drooping with the fruit. The spathes are elongate, splitting open and deciduous. The fruit is of a moderate size, oval, of a yellowish colour, and with a small quantity of rather juicy eatable pulp covering the stony seed.

On the Upper Amazon cattle eat the fruits of the Murumurú, wandering about for days in the forest to procure it. The hard stony seeds pass through their bodies undigested and become thickly scattered over the pastures adjoining the houses. They are so hard that it is almost impossible to break them, except by a very hard blow with a large hammer. The internal albumen or kernel is also excessively hard, nearly approaching to vegetable ivory. Yet pigs are very fond of these little cocoa-nuts, and on one estate on the Upper Amazon where I was staying, they had scarcely anything else to eat during a part of the year but those which had passed through the stomachs of the cows. They might constantly be seen cracking the shell with their powerful jaws, and grinding up the hard kernels, on which the teeth of few other animals could make any impression. They not only existed on this food, but in some cases got actually fat upon it. The black vultures *(Cathartes)* occasionally eat the outer covering of this and other palm fruits, when hard-pushed for food.

This tree grows on the tide-flooded lands of the Lower Amazon and on the margins of the rivers and gapós of the Upper Amazon, though it is possible that the two may be distinct species. The specimen figured is from near Pará. There are living plants in the Palm House at the Royal Kew Gardens.

A portion of a leaf is enlarged to show the spines, and a fruit is represented of the natural size.

Pl. XXXX.

ASTROCARYUM GYNACANTHUM Ht 15 Ft

145

PLATE XXXIX.
ASTROCARYUM GYNACANTHUM,
MARTIUS.
MUMBÁCA, *LINGOA GERAL.*

THIS species has a rather slender stem about fifteen feet high, covered with long, flat, black spines, arranged in regular rings and pointing downwards. The leaves are terminal, rather large and pinnate. The leaflets spread regularly in one plane, and are elongate and acute, the terminal pair being rather shorter and broader. The bases of the petioles are broadly sheathing, and are all densely spiny.

The spadices grow from the bases of the lower leaves, and are erect when in flower, but hang down with the ripe fruit, which grows in a dense cluster at the end of the long stalk which is very spiny, as is also the elongate persistent spathe. The fruit is small, ovate, of a red colour and not eatable.

This palm grows in the virgin forests of the Upper Rio Negro, and a nearly allied or perhaps identical species is common about the city of Pará.

Pl. XL.

ASTROCARYUM VULCARE Ht. 50 Ft

147

PLATE XL.
ASTROCARYUM VULGARE, *MARTIUS.*
TUCÚM, *LINGOA GERAL.*

THIS is a lofty tree, the stem growing to a height of forty or fifty feet, with a diameter of six or eight inches. It is covered with regular broad bands or rings of thickly set black spines, with narrow spaces between them. The leaves are terminal, large and regularly pinnate. The leaflets are elongate, regularly spreading and drooping. The midrib and expanded sheaths of the petioles are densely clothed with long, flat, dusky spines, having a pale expanded margin. The edges of the leaflets are also armed with fine spines.

The spadix is erect and simply branched, and is often hid among the foliage. The spathe is persistent, and the fruit is oval, of a greenish colour and not eatable.

Every part of this palm bristles with sharp spines so as to render it difficult to handle any portion of it; yet it is of great importance to the Indians, and in places where it is not indigenous, is cultivated with care in their mandiocca fields and about their houses, along with the "Pupúnha" and other fruit trees. Yet they use neither the fruit, the stem, nor the full-grown leaves. It is only the unopened leaves which they make use of to manufacture cordage, superior in fineness, strength and durability to that procured from the *Mauritia flexuosa*. They strip off the epidermis and prepare it in the same manner as described in the account of that species, but while the "mirití" is principally used for hammocks, the "tucúm" serves for bow strings, fishing-nets and other purposes where fineness, combined with strength, is required. Some of the tribes on the Upper Amazon, however, make all their hammocks of "tucúm," which renders it probable that the *Mauritia flexuosa* does not grow there.

The Brazilians of the Rio Negro and Upper Amazon make very beautiful hammocks of fine "tucúm" thread, knitted by hand into a compact web of so fine a texture as to occupy two persons three or four months in their completion. They then sell at about £3 each, and when ornamented with the feather-work borders, at double that sum. Most of them are sent as presents to Rio de Janeiro.

Dr. Martius has mistaken the species from which this cordage is manufactured, stating it to be the "Tucumá," which, though very nearly allied, is never used for the purpose. The close resemblance of the native names is probably what led to the mistake, though they are never confounded by the Brazilians.

The "tucúm" is found on the "terra firme" or dry forest land of the Amazon and Rio Negro. It is growing in the Palm House at Kew.

ASTROCARYUM TUCUMA Ht. 40 Ft.

Pl. XLI.

ASTROCARYUM TUCUMA Ht. 40 Ft.

PLATE XLI.
ASTROCARYUM TUCUMA, *MARTIUS.*
TUCUMÁ, *LINGOA GERAL.*

THIS palm is from thirty to forty feet in height, and has the stem armed with narrow rings of black spines. The leaves are terminal, rather large and regularly pinnate. The leaflets are elongate, linear and much drooping, and the midribs and petioles are very prickly. The sheathing bases of the leaf-stalks are very much swollen where they spring from the stem. The spadix grows erect from among the leaves and is simply branched. The fruit is nearly globular, of a greenish yellow colour, with a layer of yellow fleshy pulp covering the stony seed, much resembling the fruit of the Mucujá and equally esteemed for food by the Indians.

This species is very nearly allied to the last, but may readily be distinguished by its globular fruit, more drooping leaflets, less prickly habit, and the peculiar aspect of its swollen petioles. It is abundant near Pará, and is also found in the dry virgin forests of the Upper Amazon and Rio Negro.

There are young living plants in the Palm House of the Royal Botanic Gardens at Kew.

Plate II. fig. 5. represents a fruit of the natural size.

Pl. XLII.

ASTROCARYUM JAUARI Ht 40 Ft

PLATE XLII.
ASTROCARYUM JAUARI, *MARTIUS.*
JAUARÍ, *LINGOA GERAL.*

THE Jauarí has the stem rather slenderer than the Tucumá, but of about equal height, and armed with regular narrow rings of spines. The leaves are terminal and of moderate size. The leaflets are long, narrow and very much drooping, and the midribs and sheaths are thickly covered with long, flat, black spines.

The spadices are erect, simply branched, and hidden amongst the leaves. The fruit is small, oval, green, and not eatable.

The rather small dense head of foliage, combined with the prickly habit of this palm, render it altogether one of the least pleasing of the family; and the feeling is increased by its abundance in many localities, extending for miles along the river-banks to the exclusion of any other species. It is moreover one of the least useful among the larger palms, the only part which is applied to any purpose being the hard, black, oval seeds, of which the Brazilian ladies of the Upper Amazon make heads for their lace-making bobbins.

This species is unknown in the neighbourhood of Pará and on the Lower Amazon. It first occurs near Villa Nova,

153

about five hundred miles up the river, where the tidal rise and fall of the water ceases and the annual floods rise to a considerable height. From this point upwards it is very abundant, growing everywhere on the margins of the rivers, in places which are for six or eight months in the year under water. It is never found beyond the limits of the floods, and in travelling up the Rio Negro it is for hundreds of miles the only species of *Astrocaryum* met with.

Pl. XLIII.

ASTROCARYUM ACULEATUM Ht. 20 Ft

155

PLATE XLIII.
ASTROCARYUM ACULEATUM? *MEYER.*
MARAYÁ, *LINGOA GERAL.*

THIS small species has the stem from fifteen to twenty feet high and about two inches in diameter, with obscure rings of spines at irregular intervals. The leaves are terminal, rather large and regularly pinnate. The leaflets are narrow, rigid and scarcely drooping, with the terminal pair broader. The midrib and leaflets are smooth, but the bases and sheaths of the petioles are very prickly.

The spadices grow from below the leaves and are very small and simply branched. The spathes are small, ovate, swollen, erect, persistent and very prickly. The trees were not found in fruit.

This tree agrees pretty well with Dr. Martius' description of *A. aculeatum.* It grows in the virgin forest of the Upper Rio Negro.

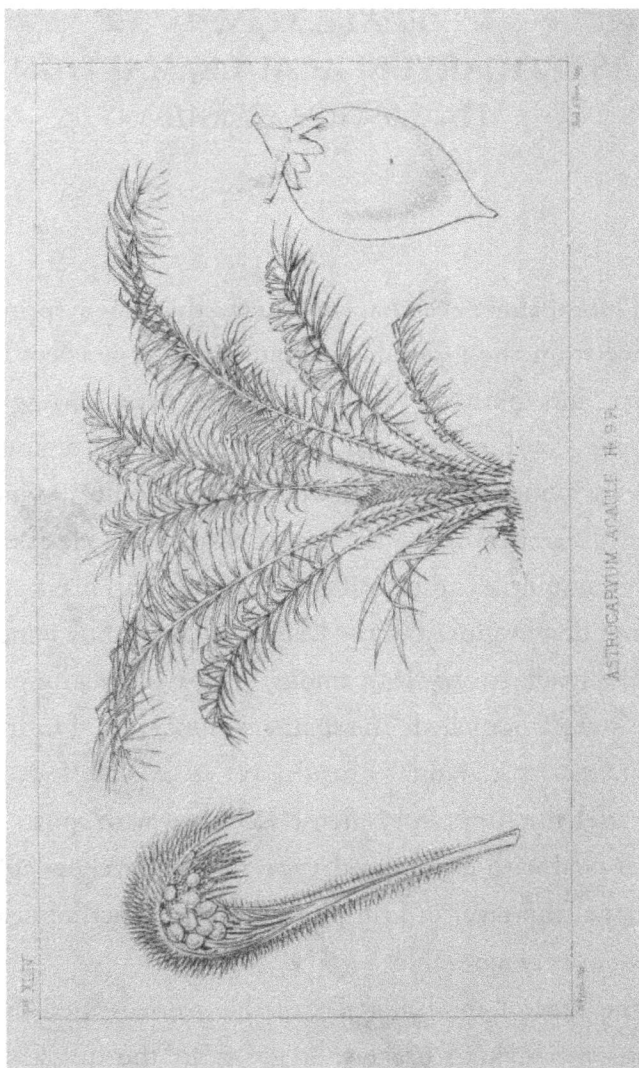

Pl. XLIV.

ASTROCARYUM ACAULE Ht 9 Ft

157

PLATE XLIV.
ASTROCARYUM ACAULE, *MARTIUS.*
IÚ, *LINGOA GERAL.*

THIS palm never has any stem, the leaves springing at once from the ground. They are eight or ten feet long, slender and pinnate. The leaflets are very narrow and drooping, and are disposed in groups of three or four, at intervals along the midrib, the separate leaflets standing out in different directions. The whole plant is exceedingly spiny, the midrib and petioles having long, flat, black spines directed downwards, and the leaflets are also spiny beneath.

The spadix grows from among the leaves on a long stalk and is simply branched. The spathe is elongate and fusiform, at first erect, but gradually bends over at the end, forming a hood over the fruit, and is densely clothed with spines. The fruit is oval with a produced apex, of a pale yellow colour, and has a thin layer of firm pulp which is sometimes eaten, but is not very agreeable.

The rind of the leaf-stalks of this palm is used by the Indians for making baskets. It grows in the dry Catinga forests of the Upper Rio Negro, often covering large tracts of ground. It has altogether a rather repulsive and inelegant appearance.

A fruit is shown on the Plate of the natural size, and a spadix reduced showing the spathe bent over it.

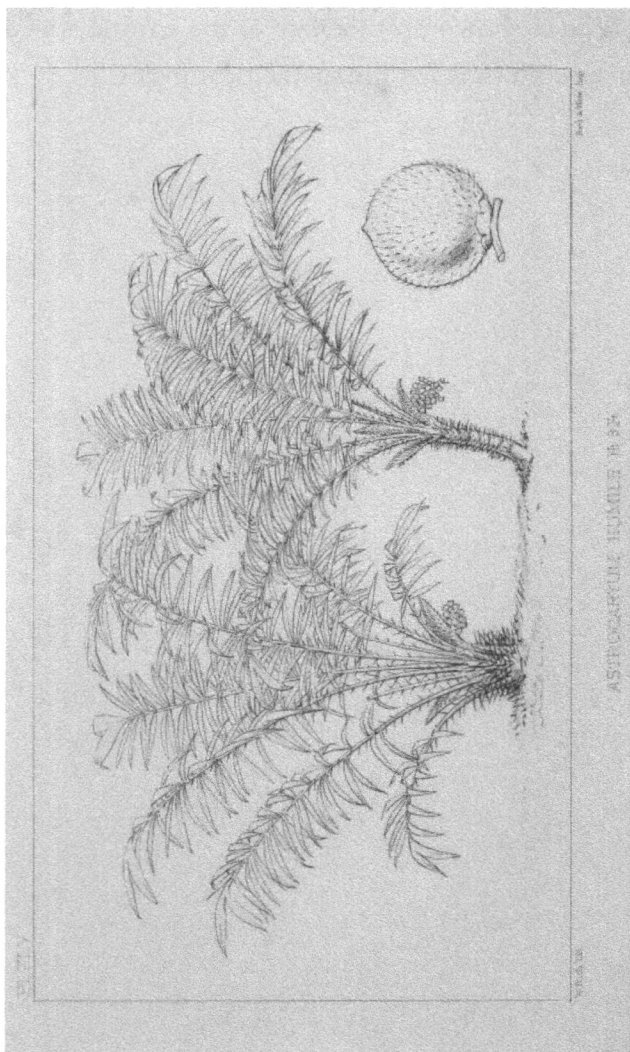

Pl. XLV.

ASTROCARYUM HUMILE. Ht 9 Ft

PLATE XLV.
ASTROCARYUM HUMILE, N. SP.
IÚ, *LINGOA GERAL.*

THIS species has a stem two or three feet high, or is altogether stemless like the last. The leaves are six or eight feet long, slender and pinnate. The leaflets are much broader than in *A. acaule,* similarly disposed in spreading groups, but not so much drooping. The midribs and petioles are armed with long, slender, cylindrical spines pointing in various directions.

The spadices grow from among the leaves and are erect and simply branched. The spathes are erect or somewhat curved over the fruit, and clothed with thickly set bristly spines. The fruit is globular, covered with scattered stiff hairs, and of an orange-red colour. It is not eatable.

This species is not uncommon in the same situations as the last. The specimen with a stem was growing in a moister part of the forest. It seems to be an undescribed species.

The stemless and short-stemmed state of this plant are shown on the Plate, and a fruit is represented of the natural size.

GENUS ATTALEA, *HUMBOLDT.*

Flowers bracteate, male and female in the same spadix, and male in another spadix, on the same or on a different tree. Spathes double, the interior one complete and woody. Male flowers with from six to twenty-four stamens and a small rudimentary pistil. Female flowers with a short style and three stigmas, and a cupshaped ring of rudimentary stamens.

The stems of these palms are generally lofty, cylindrical and smooth, but there are some stemless species. The leaves of all are very handsome, large and regularly pinnate; the petioles have the margins of the sheathing bases often more or less fibrous. The spadix grows from among the lower leaves, and is simply branched; and the fruit is ovate or oblong, and has a dry fibrous outer covering.

Sixteen species of these beautiful Palms are known, inhabiting various parts of South America, from the level of the sea to a height of 4000 feet above it. Their smooth and regularly pinnate leaves render them very suitable for thatching. One species, the *A. funifera,* produces a fibre very similar to that of the *Leopoldinia piassaba,* and the stony seeds from the same tree supply a kind of vegetable ivory.

Pl. XLVI.

ATTALEA SPECIOSA Ht 60 Ft

163

PLATE XLVI.
ATTALEA SPECIOSA, *MARTIUS.*
UAUASSÚ, *LINGOA GERAL.*

THIS noble palm has the stem fifty or sixty feet high, straight, cylindrical and nearly smooth. The leaves are very large, terminal and regularly pinnate. The leaflets are elongate, rigid, closely set together, and spreading out flat on each side of the midrib. The sheathing bases of the petioles are persistent for a greater or less distance down the stem, and in young trees down to the ground, as in the *œnocarpus batawá.*

The spadices grow from among the leaves and are large and simply branched. The fruit is of large size compared with most American palms, being about three inches long, and from this circumstance it derives its native name "Uauassú," signifying "large fruit."

The foliage of this tree is very extensively used for thatching. The young plants produce very large leaves before the stem is formed, and it is in this state that they are generally used. The unopened leaves from the centre are preferred, as, though they require some preparation, they produce a more uniform thatch. The leaf is shaken till it falls partially open, and then each leaflet is torn at the base so as to remain

164

hanging by its midrib only, which is however quite sufficient to secure it firmly. They thus hang all at right angles to the midrib of the leaf, which admits of their being laid in a very regular manner on the rafters. They are generally known as "palha branca" or "white thatch," from the pale yellow colour of the unopened leaves, and are considered the best covering for houses in places where Bussú cannot be obtained.

This species grows on the dry forest lands of the Upper Amazon. On the Rio Negro a stemless species called "Curuá" *(Attalea spectabilis)* is found and is often used for thatching. On the Lower Amazon and in the neighbourhood of Pará the *Attalea excelsa* is not uncommon. It is a handsome lofty species which grows on lands flooded at high tides, and is called by the natives Urucurí. The fruit of this tree is burnt, and the smoke is used to black the newly made india-rubber. Martius says that the fruit of the *A. speciosa* is used for this purpose, but that species is not found in the principal rubber districts, while the *A. excelsa* is abundant there.

Several species of *Attalea* are cultivated in the Palm House at Kew.

Plate III. fig. 1. is a fruit of *Attalea spectabilis* of the natural size.

GENUS MAXIMILIANA, *MARTIUS.*

Some spadices with only male flowers, others with male and female flowers on the same tree. Spathes large, complete, woody. Flowers with bracts. Male flowers with three or six stamens, and with a minute rudimentary pistil. Female flowers with a short style and three stigmas, and rudimentary stamens forming a membranaceous cup.

The stems of these magnificent Palms are tall, erect and smooth. The leaves are very large and irregularly pinnate. The bases of the petioles are persistent, often covering the stem quite down to the ground. The spathe is woody, complete, longitudinally cut and beaked. The spadices grow from among the lower leaves and are simply branched, but very densely clustered with the fruit, which is ovate, and has a dry external covering.

Only three species of this genus are known, all very handsome plants. One is a native of the West India Islands, one of Brazil, and a third is common in the Amazon district.

MAXIMILIANA REGIA

Pl. XLVII.

MAXIMILIANA REGIA Ht.80 Ft.

167

PLATE XLVII.
MAXIMILIANA REGIA, *MARTIUS.*
INAJÁ, *LINGOA GERAL.*

THIS palm has a lofty massive stem, smooth and obscurely ringed. The leaves are very large, terminal and pinnate. The leaflets are arranged in groups of three, four or five, at intervals along the midrib, from which they stand out in different directions, and are very long and drooping. The bases of the petioles are persistent a short distance down the stem, and sometimes down to the ground, even when the trees are forty or fifty feet high.

The spadices are numerous, growing from the bases of the lower leaves. They are simply branched and very densely clustered. The spathes are large, spindle-shaped, ventricose and woody, with a long beak. The fruits are elongate and beaked, with a tough, brown, outer skin, beneath which is a layer of soft fleshy pulp of an agreeable subacid flavour, covering a hard stony seed.

The leaves of this tree are truly gigantic. I have measured specimens which have been cut by the Indians fifty feet long, and these did not contain the entire petiole, nor were they of the largest size. Owing, however, to the loose irregular distribution of the leaflets, they do not produce such an effect of great size as those of the Jupati, which are more regular. The great woody spathes are used by hunters to cook meat in, as with water in them they stand the fire well. They are also used as baskets for carrying earth, and sometimes for cradles. The fruits are often eaten by the Indians, and are particularly attractive to monkeys and to some fruit-eating birds.

This magnificent palm is abundant from Pará to the Upper Amazon and the sources of the Rio Negro. It grows only in the dry virgin forest.

Young trees are growing in the Palm House at Kew, and fruit clusters and spathes are preserved in the Museum.

Plate III. fig. 3. is a view of the spathe, and fig. 2. represents a fruit, the natural size.

GENUS COCOS, *LINNÆUS.*

Female flowers less plentiful than the males, and situated below them in the same spadix. Spathe double, outer small, interior woody. Flowers with bracts. Male flowers with six stamens and a rudimentary pistil. Female flowers with three stigmas.

The stems of this genus are lofty, generally cylindrical and smooth. The leaves are large and regularly pinnate. The spadix is simply branched, and the fruit is ovate oblong, and with an outer fibrous covering.

Eighteen species of *Cocos* are known, seventeen being natives of South America, principally of Brazil, while one only, the well-known Cocoa-nut, is a native of the Old World, though it is now universally cultivated in every part of the tropics. Few species of the genus are found in the Amazon district. They appear to prefer drier and more elevated countries, some of them reaching an altitude of near 8000 feet above the sea.

Pl. XLVIII.

COCOS NUCIFERA. Ht 60 Ft

PLATE XLVIII.
COCOS NUCIFERA, *LINNŒUS.*
COQUEIRO, *PORTUGUESE.*
THE COCOA-NUT.

THE stem of this well-known palm is very smooth, seldom quite erect, and often much thicker at the bottom. The leaves are large, terminal and regularly pinnate. The leaflets are rigid, and spread out very flat on each side of the midrib. From the sheathing bases of the petioles grows a compact fibrous material resembling in texture the spathe of the Bussu.

The spadices are produced from among the leaves, and are large and simply branched. The fruits are very large, and have a dense fibrous external covering over the well-known cocoa-nut.

This tree is not a native of South America, but as it is generally cultivated in every part of the tropics, I have given a figure of it. Its peculiar characteristic is the rigidity of its leaves, which curve or droop very slightly, and the leaflets spread out with remarkable flatness and regularity. The stem also is rather massive in accordance with the immense weight of fruit which it produces, and the whole tree, though exceedingly handsome, has not that light and feathery appearance which it is often represented as possessing. It is not impossible, however, that it may have acquired by its naturalization in America an aspect differing somewhat from its characteristic features when growing on the sea-shore, on the coral islands of India and the Pacific.

There it is of the greatest utility to man. It supplies food and drink and oil. Its fibres are woven into cordage and matting, and it even furnished animal as well as vegetable food, herds of swine being fed and fattened entirely on its fruit.

On the banks of the Amazon, on the contrary, we see at once that it is in a foreign land. It flourishes indeed with great luxuriance, but no part of it is applied to any useful purpose, the fruit only being consumed as an occasional luxury. In the towns and larger villages where the Portuguese have settled it has been planted, but among the Indians of the interior it is still quite unknown.

LIST OF THE PALMS DESCRIBED IN THIS WORK, WITH THEIR NATIVE NAMES AND USES.

Botanical Name.	Native Name.	Uses.
Leopoldinia		
pulchra	Jará	Stem used for fencing, rafters, &c.
major	Jará assú	Fruit for making salt.
piassaba	Piassába	Fibre for cordage, brooms, &c.; leaves for thatching; fruit eatable.
Euterpe		
oleracea	Assaí	Fruit for making a drink; stem for rafters, &c.
catinga	Assaí de Catinga.	Fruit for making a drink.
œnocarpus		
baccaba	Baccába	Fruit makes a drink and oil; leaves for thatching.
batawá	Patawá	Fruit makes a drink; spinous processes used for making arrows.
disticha	Baccába	Leaves for thatching.
minor	Baccába miri	Fruit makes a drink.
Iriartea		
exorhiza	Pashiúba	Stem split for floors and ceilings, &c.; air-roots for graters.
ventricosa	Pashiúba barriguda	Stem split for lances, harpoons, floors, &c.; swollen part of stem for canoes.
setigera	Pashiúba miri	Stem hollowed for making blow-tubes or Gravatánas.

174

Raphia		
tædigera	Jupatí	Leaf-stalks split for making boxes, partitioning houses, doors, &c.
Mauritia		
flexuosa	Mirutí	Fruit makes a drink; fibres of leaves are twisted into string for hammocks, &c.; leaf-stalks as the last.
aculeata	Caranaí	Fruit makes a drink.
gracilis	Caranaí	Fruit makes a drink.
pumila	Caranaí	Not known.
caraná	Caraná	Leaves good for thatch; leaf-stalks used as those of *Raphia tædigera.*
Lepidocaryum		
tenue	Caranaí do Mato.	None.

Botanical Name.	Native Name.	Uses.
Geonoma		
multiflora	Ubimrána	These species and others allied all have the leaves more or less used for thatching.
paniculigera	Ubim de Cotiwiya	
rectifolia	Ubimrána	
Manicaria		
saccifera	Bussú	Leaves the best for thatching; spathe for caps, wrappers, &c.
Desmoncus		
macroacanthus.	Jacitára	Bark makes "tipitis" or elastic cylinders for squeezing the grated mandiocca.
Bactris		
pectinata	Iú	These little prickly palms seem not to be applied to any particular uses.
n.s	Marayarána	
elatior	Marayarána	
n.s	Unknown	
macrocarpa	Iú	
tenuis	Iú	
simplicifrons	Iú	
maraja	Marajá	Fruit eatable.
integrifolia	Iú	None.

Guilielma		
speciosa	Pupúnha	Fruit very good and nutritious; wood very hard, black and durable.
Acrocomia		
lasiospatha	Mucujá	Fruit eatable.
Astrocaryum		
murumurú	Murumurú	Cattle eat fruit.
gynacanthum	Mumbáca	None.
vulgare	Tucúm	Leaf-fibres for cordage.
tucumá	Tucumá	Fruit eatable.
jauarí	Jauarí	Nuts for lace-bobbin heads.
aculeatum	Marayá	None. Others with the same name have eatable fruit.
acaule	Iú	Bark of leaf-stalks for baskets.
humile	Iú	Fruit eatable.
Attalea		
speciosa	Uauassú	Leaves for thatch.
excelsa	Urucurí	Fruit burnt for smoking rubber.
spectabilis	Curuá	Leaves for thatch.
Maximiliana		
regia	Inajá	Fruit eatable.
Cocos		
nucifera	Coqueiro	The Cocoa-nut; fruit eatable.

The genera of Palms found in America are thirty-six in number. Thirty-two of these are entirely confined to it, while only four are common to the Old and New Worlds, as shown in the following list:—

LIST OF THE AMERICAN GENERA OF PALMS.

Name of Genus.	No. of species mentioned in this Work.	Species found in America.	Species of American Genera in the Old World.
Chamedorea	0	23	0
Hyospathe	0	1	0
Morenia	0	2	0
Kunthia	0	1	0
Leopoldinia	3	4	0
Euterpe	3	12	0
œnocarpus	4	6	0
Oreodoxia	0	6	0
Reinhardtia	0	1	0
Iriartea	4	9	0
Ceroxylon	0	3	0
Raphia	1	1	2
Mauritia	7	8	0
Lepidocaryum	1	2	0
Geonoma	3	33	0
Manicaria	1	1	0
Copernicia	0	6	0
Brahea	0	2	0
Sabal	0	9	0
Trithrinax	0	2	0
Chamærops	0	2	6
Thrinax	0	8	0
Desmoncus	1	14	0
Bactris	9	46	0
Guilielma	1	3	0
Martinezia	0	4	0

Acrocomia	1	8	0
Astrocaryum	8	17	0
Elœis	0	1	1
Attalea	3	16	0
Maximiliana	1	3	0
Orbignia	0	3	0
Syagrus	0	5	0
Diplothemium	0	5	0
Jubæa	0	1	0
Cocos	1	17	1
Totals	52	285	10